Future Agriculture—
"Viksit Bharat"

Future Agriculture— “Viksit Bharat”

Editors

Dr. Kirit Shelat
Former Principal Secretary (Agriculture), Government of Gujarat & Executive Chairman, NCCSD, Ahmedabad, Gujarat, India

Prof. Shrikant Kalamkar
Director Professor, Agro-Economic Research Centre (MoA & FW, GoI), Sardar Patel University, Vallabh Vidyanagar, Anand, Gujarat, India

Dr. A.R. Pathak
Former Vice-Chancellor, Junagadh Agricultural University (JAU) & Navsari Agricultural University, Navsari, Gujarat, India

Dr. Odemari Mbuya
Professor of Agricultural Sciences, Director of the Center for Water Resources at Florida Agricultural and Mechanical University (FAMU), USA

ALLIED PUBLISHERS PVT. LTD.

Noida • Mumbai • Kolkata • Chennai • Bangalore

ALLIED PUBLISHERS PRIVATE LIMITED

D-5, Sector-2, **Noida**–201 301
Ph. Nos.: 0120-4320295/2542557/4352866
E-mail: delhi.books@alliedpublishers.com

17 Chittaranjan Avenue, **Kolkata**–700072
Ph.: 033-22129618 • E-mail: cal.books@alliedpublishers.com

15 J.N. Heredia Marg, Ballard Estate, **Mumbai**–400001
Ph.: 022-42126969 • E-mail: mumbai.books@alliedpublishers.com

No. 25/10, Commander-in-Chief Road, Ethiraj Lane (Next to Post Office)
Egmore, **Chennai**–600008
Ph.: 044-28223938 • E-mail: chennai.books@alliedpublishers.com

P.B. No. 9932, No. 15, 3rd Floor (Next to Vijaya Bank), 5th Cross, Gandhinagar, Karnataka, **Bangalore**–560009. Ph.: 080-41530285/22386239
E-mail: bngl.books@alliedpublishers.com

Website: www.alliedpublishers.com

First Edition: 2024

ISBN: 978-93-89934-43-4

Editors would be happy to receive comments and sharing of experiences from readers.

Citation: Shelat, Kirit, Shrikant Kalamkar, A.R. Pathak and Odemari Mbuya (2024). (Eds), "Future Agriculture—'Viksit Bharat', Allied Publishers Pvt. Limited, Noida

Disclaimers: Editors take no responsibility of the statement made by the authors in their papers. It is solely the personal opinion of respective paper author/s.

AERC-Publication Officer and Circulation In-charge: Dr. Deep K. Patel, Res. and Ref. Assistant (Lib).

Published by Sunil Sachdev and printed by Ravi Sachdev
at Allied Publishers Pvt. Ltd., D-5, Sector-2, Noida–201 301

डॉ. पी. के. मिश्र
प्रधान मंत्री के प्रधान सचिव
Dr. P. K. Mishra
Principal Secretary to Prime Minister

सत्यमेव जयते

प्रधान मंत्री कार्यालय
साउथ ब्लॉक, नई दिल्ली - 110 011
Prime Minister's Office
South Block, New Delhi - 110 011

Foreword

This book titled *Future Agriculture – "VIKSIT BHARAT"* is a compilation of articles focusing on important aspects of agriculture. Agriculture today faces challenges of climate change, resource scarcity and need for sustainability. In this book experts and practitioners have attempted to address some of these issues. The topics include carbon sequestration, water scarcity and empowerment of women in agriculture, capturing their vital role in sustainable development.

During the last decade Prime Minister Shri Narendra Modi has taken several measures to accelerate transformation in the agriculture sector with an objective to empower our farmers and improve their income and living standards. The book highlights several initiatives taken by the Government of India.

The writers have suggested strategies to make agriculture more sustainable, productive and remunerative. The book brings out key challenges in agriculture sector as well as opportunities and the way forward. It highlights the importance of modern technology and a more effective extension strategy suited to the present-day needs. Success stories from different parts of the country are also described. These can be replicated so as to enhance the agricultural outcomes.

Dr. Kirit Shelat and others who have contributed to the book have studied and experienced development issues during their distinguished careers. I personally know most of them. I compliment their efforts. Indeed, this book is a testament to the collaborative efforted needed to navigate the challenges of today's agriculture and unlock its full potential.

This book is an important contribution which will be useful to policy makers, practitioners and students.

P.K. Mishra

Telephone : +91-11-2301 3040 Fax : +91-11-2301 7475 e-mail : pkmishra.pmo@gov.in

Prof. Dr. Niranjan Patel

Vice Chancellor,
Sardar Patel University,
Vallabh Vidyanagar
388120, Anand,
Gujarat

June 15, 2024

Message

Agriculture sector continues to play a vital role in the country's socio-economic development and progress of our nation. Despite the relatively higher growth of the non-agriculture sector, it is the agriculture sector that remains the principal source of employment, with 45 per cent of the workforce engaged in agricultural and allied activities. Serious efforts were then undertaken by the Union and States for wider dissemination of Green Revolution technology and towards diversification in favor of high value crops. During the seven decades since Independence, the agriculture sector witnessed phases of high and low growth. Soon after the spread of modern technology in Indian agriculture, there was much discussion on the first-generation problems of the Green Revolution. These problems became more severe over time and some were aggravated by government policies. These challenges are related to sustainability of natural resources, efficiency, and the plight of farmers with small holdings, food safety, profitability, fiscal effect and equity. Agriculture policy should address these challenges during AmritKaal. The four 'Amrit' pillars of the 'AmritKaal' are our women power, our youth power, our agricultural strength, and the empowerment of our poor and middle-class, will undoubtedly elevate Bharat to new heights and build a developed nation.

Agro-Economic Research Centre (AERC) of our University and National Council for Climate Change Sustainable Development & Public Leadership (NCCSD), Ahmedabad has started jointly working on the emerging issues in Indian agriculture and strategies to adopt during Amritkaal period. In fact, AERC since its inception has been working on policy formulation related agriculture and rural development in general and irrigation water in particular. NCCSD is known for its think tank output. I must compliment both these Institutions, viz. AERC and NCCSD for this outcome. I thank policy makers, academicians, stakeholders and all related personal who have contributed in this book.

I congratulate the editors of book, Dr. Kirit Shelat, Dr. Shrikant Kalamkar, Dr. A.R. Pathak and Dr. Odemari Mbuya for bringing this valuable book.

June 10, 2024

Prof. Dr. Niranjan Patel

Preface

The relentless march of technological progress has indelibly transformed the agricultural landscape of our nation. We have witnessed a remarkable transition from an era of food scarcity to one of abundance, a testament to the transformative power of innovative technologies. However, amidst these achievements, the unintended consequences of improper use of technologies like the loss of biodiversity including soil biota, the deterioration of natural resource base, reduced resource use efficiency, increased cost of cultivation, emergence of new diseases and pests have augmented the woes of the farmer in terms of increased cost of cultivation and low productivity.

Soil quality, salinity ingression and the declining quality and content of irrigation water, market intelligence, weather forecast are among the pressing challenges that can be addressed through the judicious and informed use and management of technologies tailored to local contexts. It is here that innovation-based technologies hold immense potential to forge a sustainable and prosperous future for Indian agriculture.

Technologies like precision farming, biotechnology, artificial intelligence and remote sensing are poised to revolutionize the sector when they are customized to specific situations, optimizing water usage, managing nutrient cycles, detecting diseases early and maximizing yield potential – innovations that can directly contribute to enhancing farm income and ensuring long-term sustainability.

The Government of India recognizes the importance of this symbiotic relationship between farmers and technology. It actively promotes field monitoring by drones, weather forecasting, decision-making guided by satellite imagery, and data-driven information dissemination for different activities. This holistic approach aims to empower farmers with the tools and knowledge necessary to navigate the complexities of modern agriculture while safeguarding the environment.

This book serves as a comprehensive guide to navigating the intricate landscape of future agriculture technologies. This has consummated wealth of information on the latest advancements, their practical applications and their potential to address the unique challenges faced by Indian farmers. From precision farming techniques that optimize resource utilization to biotechnological innovations that enhance crop resilience, this informational treatise explores a wide array of cutting-edge solutions. Moreover, it underscores the vital role of traditional wisdom, such as agroforestry and natural farming, in creating a holistic approach to shaping a sustainable agricultural future.

The transformative journey of agriculture cannot be accomplished without acknowledging the pivotal role of our farmers, particularly the rural women who form the backbone of Indian agriculture. This book emphasizes the moral and strategic imperative of empowering these vital stakeholders to unleash their immense potential in shaping the future of agriculture.

By striking a harmonious balance between technological advancements and environmental stewardship, we can unlock a future where agriculture not only thrives but also contributes to the preservation of our planet's precious resources. It is a path that demands continuous innovation, adaptability and a deep respect for the intricate web of life that sustains us all. It is our hope that this book will serve as a valuable resource for policymakers, researchers, agricultural professionals and farmers alike, inspiring a collective effort to harness the power of innovation while preserving the wisdom of our agricultural heritage.

Kirit N. Shelat, Shrikant S. Kalamkar,
A.R. Pathak & Odemari Mbuya

About the Editors

Dr. Kirit Shelat *(IAS Retired)*

Former Principal Secretary, Government of Gujarat
&
Executive Chairman, NCCSD, Ahmedabad

Dr. Kirit Shelat is Doctorate in Philosophy with Public Administration. He was awarded a Degree of D.Litt.-Doctorate of Science by Junagadh Agricultural University, Junagadh, Gujarat (India) for his outstanding contribution in promoting 'Climate Smart Agriculture and Building Climate Smart Farmers'. He had long spell of his carrier in Indian Administrative Service. He has hand into introduction in 'New Extension Management- Krishi Mahotsav' approach in Gujarat as Principal Secretary-Agriculture which doubles the income of farmers. He has designed and implemented large scale projects for poor families, farmers and micro entrepreneurs and remote rural areas. He has authored more than 20 books related to agricultural and rural development and related to impact of climate change and ways to meet that challenge at local level-village level. He is Executive Chairman of National Council for Climate Change, Sustainable Development and Public Leadership (NCCSD).

Dr. Kirit Shelat has written biography of Pujya Pramukh Swami Maharaj- "YUG PURUSH, PUJYA MAHANT SWAMI MAHARAJ- a life dedicated to thers". This is published in six languages with nine editions. He has also written biography of PujyaMahant Swami Maharaj- "Mahan Rushi Mahant Swami Maharaj"-June 2019.

He was member of sub-committee set us by Planning Commission of India on enhancing preparedness for Climate Change and has his hand in introduction of NICRA-National Initiatives for Climate Resilient Agriculture. He was member of expert committee of Government of Gujarat on Economic Revival in Arena Covid Pandemic". Along with Prof Mbuya of FAMU USA, he introduced the concept of "Building Climate Smart Farmers'. His latest publication includes "Atmanirbhar Bharat – "self-reliant and climate smart farmers – Roadmap for Agriculture 2020-30 India" and "Atmanirbhar Farmers – Gujarat – Roadmap 2030". He was of also member of Selection Committee for Prestigious Padma Award for year 2023–24.

Dr. Shrikant S. Kalamkar

Director and Professor
Agro-Economic Research Centre
(Ministry of Agriculture & Farmer Welfare, Govt. of India)
H.M. Patel Institute of Rural Development,
Sardar Patel University, Vallabh Vidyanagar 388120

Dr. S.S. Kalamkar is presently working as Director and Professor in Agro-Economic Research Centre (AERC) of the Ministry of Agriculture and Farmers Welfare, Government of India located at Sardar Patel University, Vallabh Vidyanagar, Anand, Gujarat since July 2012. AERC is established and fully supported by the Ministry of Agriculture and Farmers Welfare, Govt. of India. Before joining AERC, he has worked with Gokhale Institute of Politics and Economics (GIPE), Deemed University, Pune (from 2002 to 2012), and ICAR-National Bureau of Soil Survey and Land Use Planning, Nagpur (from 2000 to 2002).

He has made important and useful contributions to the subject of agricultural economics. He has published more than 155 research papers/articles in reputed journals. He has 13 books at his credit. Besides, he has participated in more than 175 national conferences/workshops/seminars. Dr. Kalamkar has completed about 64 research projects/studies, mainly for the Ministry of Agriculture, Govt. of India; the Planning Commission, Govt. of India, Government of Gujarat and NABARD, Mumbai; NDDB, Anand. He has submitted notes on topics of national importance as policy briefs and policy alerts (coordinated and published by CMA, IIM, Ahmedabad) from time to time on to the Hon'ble Prime Minister's Office, the Ministry of Agriculture and Farmers Welfare, GOI & others for consideration in policy formulations.

He is the recipient of Sardar Patel Research Award 2017–18, Shri Dadabhai Navroji Prize (2017–18), SOED-Ludhiana Best Paper Award'; Prof. V.S. Vyas Felicitation award 2014–16 & 2016–18 for the best report completed at AERC. He is also the recipient of the University Gold Medal and Cash Prize Award in M.Sc. (Agri.).

Besides significant contribution in research, Dr. Kalamkar has also worked as Registrar (Officiating) of the GIPE (Deemed University), Pune from August 2007 to March 2009 as well as Registrar (In-charge) of Sardar Patel University, Vallabh Vidyanagar during March 14, 2015 to April 4, 2015. He is elected as a President of Gujarat Economic Association, Anand (2024–2025).

Dr. A.R. Pathak

Former Vice Chancellor, Junagadh Agricultural University, Junagadh & Navsari Agricultural University, Navsari

Dr. A.R. Pathak Former Vice Chancellor, N.A.U. Navsari (2010–2014) and J.A.U, Junagadh(2014–2019) Agriculture University has excellent report among the scientist and farming community by virtue of his hard work and contribution helped the rural farmers as well as people's organisation. He was Director of Research, AAU, Anand for 5 years. He as a Plant Breeder worked on development of improved high yielding, pest and disease resistance varieties of pulses, castor, mustard and rice. In all a total of 29 varieties of different crops were developed and released for farmers by central/state variety release committee. Dr. Pathak as a Vice Chancellor of JAU and NAU established various colleges in the discipline of agriculture, agriculture engineering, agriculture biotechnology, and horticulture with all facilities and faculties. The colleges and research centers were strengthened with man power, equipment and other required infrastructure. NAU & JAU with all colleges were accredited by ICAR under his leadership. Important research work on value adition of banana psuedostem, research on value addition and post-harvest, sugarcane, rice, cotton, mango, floriculture, groundnut and vegetable were strengthened to help farming community. Mass production of tissue culture plant lets of sugarcane, banana, seed production of improved varieties of rice, groundnut, wheat and other crops as well as bio fertilizer and bio pesticide production were done and provided to farmers at reasonable price. He was constantly working and performing activities related to agriculture and allied fields for the benefit of farmers and rural peoples. He has served in several committees at National level viz; Accreditation committee of various universities, selection committee ASRB,/universities, BoM of CAU Imphal, Appeal committee, QRTs as Chairman or member including Chairman of Empowered committee of NASF. He also served as President IAUA & GAAS. He visited IRRI, Philippines, USA, Sweden, Bangkok, Thailand and Dubai for research and education purpose. He has received 11 awards in the field of agriculture including Sardar Patel Research award by Government of Gujarat, Cereal award by GAAS Lifetime Achievement award by Agriculture Today, GSIA and best Vice Chancellor by AIASUA, ICAR.

Dr. Odemari Stephen Mbuya

*Professor of Agricultural Sciences,
Director of the Center for Water Resources at Florida Agricultural and Mechanical University (FAMU), US*

Dr. Odemari Stephen Mbuya is a professor of Agricultural Sciences, Director of the Center for Water Resources at Florida Agricultural and Mechanical University (FAMU, 1601 S. Martin Luther King Jr Blvd, Tallahassee, FL 32307), and a Courtesy Professor of Agronomy at the University of Florida/IFAS, where he teaches Statistical Research Methods, Plant and Soil Sciences. With broad and extensive knowledge in agriculture, life sciences and sustainable ecosystems for the past thirty years, Dr. Mbuya leads the faculty effort in integrating all concepts of sustainability in innovative research and teaching at FAMU. With his extensive education and research career he has trained undergraduate and graduate students as well as postdoctoral researchers. He has worked as a research scientist in Tanzania, a visiting researcher at the Centro Internacional de Agricultura Tropical (CIAT) in Colombia, a consultant and volunteer representing the United States Department of Agriculture (USDA) and FAMU in South Africa and India, and a delegate of Institutions of Higher Learning from United States to the Netherlands.

Dr. P.K. Mishra

Principal Secretary to Prime Minister
Government of India, New Delhi

Dr. P.K. Mishra, at present Principal Secretary to Prime Minister of India, has, as a former member of the Indian Administrative Service, varied work experience (holding senior positions) in field organizations as well as at the policy-making levels of the government. He is a 1972 batch Indian Administrative Service officer belonging to Gujarat cadre. He headed a Task Force of the Union Home Ministry, during 2012–13, to review the Disaster Management Act, 2005. He also chaired a committee constituted by the Union Agriculture Ministry to review the working of crop insurance schemes. He was also on the International Advisory Group of the World Bank to prepare a Disaster Recovery Framework at the global level. During 2001–2004, Mishra has also served as the principal secretary to Narendra Modi, when he was the Chief Minister of Gujarat.

As Secretary to Government of India, Ministry of Agriculture during 2006–08, Dr. Mishra was actively involved in path-breaking national initiatives such as the National Agriculture Development Programme (RKVY) and the National Food Security Mission (NFSM). He functioned during 2001–2004 as Chief Executive Officer of the Gujarat State Disaster Management Authority (GSDMA) created after the Kutch earthquake of 26 January 2001. He played a crucial role in shaping the GSDMA into a vibrant, dynamic and innovative organization, which received a number of prestigious international awards – including those of the United Nations, and the Commonwealth Association for Public Administration and management (CAPAM) – for outstanding work in disaster recovery and management. He published a book entitled 'The Kutch Earthquake 2001: Recollections, Lessons and Insights in 2005'. He has written a book entitled 'Agricultural Risk Insurance and Income' published from UK in 1996. He has edited a book on Agricultural Insurance for the Asian Productivity Organization, Tokyo. He has published a number of papers in national and international journals, and has been invited to many international conferences as resource person and to present papers. He has a Ph.D in Economics/Development Studies from the University of Sussex, UK. He completed M.A. degree in Development Economics at the University of Sussex in 1990, when his academic performance was rated outstanding by the university. He did his M.A. in Economics with a first class at the Delhi School of Economics in 1972.

About the Paper Authors

Dr. Vijay Kumar Saraswat

Member, NITI Aayog, and Former Chancellor, Jawaharlal Nehru University, Delhi

Dr Vijay Kumar Saraswat is a distinguished scientist with vast experience in defence research—in both basic and applied sciences—spanning several decades. He retired as Secretary, Defence Research and Development Organisation (DRDO), after decades of government service. He holds a PhD from Osmania University and an ME degree from the Indian Institute of Science. Dr Saraswat has had an illustrious career and is credited with the indigenous development of missiles such as *Prithvi*, *Dhanush*, *Prahaar*, and *Agni-5;* the two-tiered Ballistic Missile Defence system; the initial operational clearance of Light Combat Aircraft *Tejas;* and nuclear submarine *INS Arihant*. He is the recipient of many awards, including the Padma Shri (1998) and Padma Bhushan (2013). He has been conferred an honorary doctorate by more than 25 universities, most recently in 2018 by JamiaHamdard.

Dr. C.K. Timbadia

Vice Chancellor, Gujarat Natural Farming Science University, Halol, Panchmahal, Gujarat

Dr. Chandulal Karamshibhai Timbadia is the first Vice Chancellor of new established Gujarat Natural Farming Science University at Halol, Panchmahal, Gujarat by the Government of Gujarat. He has worked as Director of Extension Education, NAU, Navsari; Senior Scientist and Head, Krishi Vigyan Kendra, NAU, Navsari; Project Director of ATMA Project, Tapi. He was nnominated as committee member for preparing syllabus of Natural farming by ICAR, New Delhi and NCERT. He was instrumental as a committee member in preparing syllabus of Natural farming by Government of Gujarat. He has received many awards for his extensive work.

Dr. (Ms) Neelam Patel

Senior Advisor (Ag), NITI Aayog, GOI

Dr. Neelam Patel has been serving as a Senior Advisor (Agriculture and Allied Sectors) at National Institute of Transforming India (NITI) Aayog, the policy Think Tank of the Government of India.

Dr. Meenesh Shah

Chairman & Managing Director, National Dairy Development Board, Anand

Dr. Meenesh Shah has an illustrious career of more than 37 years in Dairy sector and has made immense contribution in the field of Research & Development, Product & Process Development, Project Appraisal & Finance, Cooperative Services, Project Management and Strategic Decision Making. Dr. Shah is also Chairman of MDFVPL, IIL, IDMC, NDDB Dairy Services & NDDB MRIDA, furthering the objectives of NDDB and catering to various value chain of our animal husbandry and dairying sector from animal & human health, dairy machineries, institution building, marketing of milk & milk products to manure management. He is also in the Governing Boards of many other premier institutions of the country. He is Member Secretary of Indian National Committee of International Dairy Federation and also a member of the Standing Committee on Dairy Policies and Economics of IDF. He spearheaded successful organisation of the World's largest Dairy Conference IDF-WDS 2022 in India in September 2022. He had also ensured successful implementation of the World Bank funded Central Sector Scheme 'NDP Phase I'.

Dr. Rajendra Shende

Former Director UNEP. Coordinating Lead Author of Nobel Peace Prize-winning IPCC, IIT Alumni, Founder Director of Green TERRE Foundation Former Director UNEP

Dr Rajendra Shende, an alumnus of IIT Bombay and former Director of United Nations Environment Programme (UNEP) in Paris. Before joining UN, he had varied experience as senior manager in private sector including Tatas in India. He led the global programme at UNEP to save the Ozone Layer. That programme is till date is termed by UN as uniquely successful environmental programme in the history of UN. As Coordinating Lead Author of Nobel Peace Prize winning Intergovernmental Panel on Climate change (IPCC) in 2007 that won Nobel Peace Prize he has been awarded citation of Nobel Peace Prize in 2007. He is Founder Director of not-for-profit -Green TERRE Foundation, It is action-platform for seeking sustainable solutions to our developmental imperatives. Its Global Network of Universities for localising SDGs and Net Zero in the campus is spread in 12 countries. He serves as member of Committee of India's Education Ministry for National Institutional Ranking Framework (NIRF). Winner of numerous awards including two from USA government, one from Harvard University's JFK school of governance, one from UNEP, one from UK Industry and numerous from India including Life time Achievement from BAMU.

Dr. V.P. Chovatia

Vice Chancellor of Junagadh Agricultural University, Junagadh (Gujarat)

Dr. V.P. Chovatia is presently the Vice Chancellor of Junagadh Agricultural University, Junagadh (Gujarat). He has served as a Director of Research and Dean PG Studies, Junagadh Agricultural University, Junagadh.

Dr. V.V. Sadamate

Agricultural Extension Specialist & Former Adviser Agriculture. Planning Commission (Now NITI Aayog), New Delhi

Dr. V.V. Sadamate has served the Indian extension system in various capacities during his more than 3-decade stint with the Ministry of Agriculture and the Planning Commission. He also served as a consultant with the World Bank supported Neeranchal Project to develop convergence strategies and with the FAO, India in developing the Country Programme Framework. Still actively associated with GO/NGO/ Company-Corporates' outreach programmes in advisory capacities. Contributing in Ag. Extension academics, practices and applications in various fields/tasks.

Dr. K.D. Nayak

Former Director General (Micro-electronics Devices, Computational Systems and Cyber Security), DRDO, Ministry of Defense

Dr. K.D. Nayak obtained Bachelor of Engineering from UVCE Bangalore and Doctrate from IIT Kanpur. He joined DRDO in 1984 and served in various positions as CEO SITAR, Director ANURAG and CC (R&D) (MED & MIST). He is recipient of numerous accolades and awards, IGMDP Award in 1989, IETE-IRSI (83) Award in 1999, Scientist of the year Award in 2006 and Technology Leadership Award in 2013. He has been the chairman and member of various national cyber policy committee, national cyber infrastructure indigenization committee, and Technical Advisory board of SAMEER, CDAC etc. He has more than 40 technical papers in national and international journals.

Dr. A. Arunachalam

Director, ICAR-Central Agroforestry Research Institute, Jhansi 284003, Uttar Pradesh

Dr. A. Arunachalam is presently the Director of ICAR-Central Agroforestry Research Institute, Jhansi, Uttar Pradesh, and Project Coordinator of (i) DST supported Task Force of Himalayan Agriculture under the National Mission on Sustaining Himalayan Ecosystems and (ii) ICAR supported All India Coordinated Research Project on Agroforestry. He is designated Chairman of the Task Force on Agroforestry constituted by NITI Ayog, Govt. of India. Bangladesh. He has completed more than 35 research projects and published over 400 articles in journals/ books and edited several books. Dr. Arunachalam has more than 30 years of teaching, research and extension, and research management in the areas of botany, forestry, ecology, biodiversity and agroforestry. He is a Fellow of 32 national academies & professional societies and is a recipient of 25 national/international awards.

Dr. Suresh Acharya

Former Director of Research, SDAU, SK Nagar & R&D Vertical, AGROCEL Industries Pvt. Ltd. Koday, Mandvi, Kachchh

Dr. Suresh Acharya is an expert in Crop Improvement and multi-disciplinary Research Management. After serving in five different SAUs of the country, he hund his boots in 2017 as Director of Research, SDAU, SK Nagar, Gujarat. At present, he is working in AGROCEL Industries Pvt. Ltd. Koday, Mandvi, Kachchh, Gujarat. He has bred 19 versatile varieties (wheat, cotton and pulses), 81 R Lines and 110 A lines in pigeonpea. He guided 12 M.Sc. and 09 PhD Students. He handled 15 diversified projects as Principal Investigator; and organized 16 national level programs as Organizing Secretary/Nodal officer/Course Coordinator. He has been a member of 17 important committees that also included Extant Varietal Identification Committee, PPV & FRA, New Delhi. He was in the editorial panel of different National Journals and was Editor in Chief of GAU Journal Research.

Dr. D.K. Varu

Professor & Head, Department of Horticulture, Junagadh Agricultural University, Junagadh (Guj.)

Dr. Devsi Karnabhai Varu, M.Sc. (Agri.) & Ph.D. (Horticulture) has more than 28 years of professional experience in Agricultural Education, Research, Extension and administration. He has guided 23 M.Sc. (Horti.) and 10 Ph.D. students. He published more than 97 research papers International journals, 12 in national journals, while, more than 100 scientific publications including popular articles. He has received more than nine awards and appreciations from various organizations including felicitation with Best Teacher Award by ICAR. He has contributed in development of more than 45 technologies and release of two new varieties like GJCA-1 in custard apple and GJP-1 in papaya. He also contributed for the endorsement of seven varieties in different horticultural crops. He also identified one genotype in black jamun. He has coordinatedmore than 20 research projects at JAU. He also contributed to achieve the Geographical Indication (GI) for the Kesar variety of mango as "GirKesar Mango".

Dr. A.K. Handa

Principal Scientist (Forestry/Agroforestry) ICAR-Central Agroforestry Research Institute, Jhansi

He is serving as Principal Scientist (Forestry/Agroforestry) at ICAR-Central Agroforestry Research Institute, Jhansi 284003, Uttar Pradesh. His areas of interest are Agroforestry & Policy, Tree Improvement, Treecrop Interaction.

Dr. Sanjay Deshmukh

Professor of Life Sciences; Former Vice-Chancellor, University of Mumbai, Mumbai

Prof. Sanjay Deshmukh currently holds the esteemed position of Professor of Life Sciences at Mumbai University, and also serves as an Adjunct Professor at the Institute of Chemical Technology, Mumbai. He is a distinguished academic thought leader in India, renowned for his exceptional leadership qualities, commitment to organizational success, and extensive educational background. As the youngest full-time Vice-Chancellor of the University of Mumbai, Prof. Deshmukh spearheaded its remarkable transformation characterized by transparency and the implementation of global best practices. His focus on skill development and employability aimed to equip students for the professional world. Prof. Deshmukh's academic prowess is reflected in his diverse educational qualifications, encompassing degrees in Science and Law. This multidisciplinary approach fuels his innovative thinking and problem-solving abilities, contributing to his exceptional achievements. He has been bestowed with two honorary DSc Degrees.

Prof. R. Gopichandran

Professor, NTPC School of Business, Noida, U.P. India

Prof. R. Gopichandran holds two doctoral degrees in the areas of microbial and chemical ecology, followed by a degree in law. Most of his 35 years of work has been on facilitating compliance with the Montreal Protocol, working closely with the Compliance Assistance Programme, Ozon Action programme of the UNEP at the national, regional and global levels. Prior to his professorship at the NTPC School of Business, he served also as Director, Vigyan Prasar, an autonomous organization of the DST, Government of India, Principal Research Scientist – Environment & Climate Change at the Gujarat Energy Research Management Institute, Gandhinagar and up to level SG at the Centre for Environment Education, Ahmedabad.

Dr. Suresh Ramanan S.

Scientist, M.Sc. (Agroforestry) ICAR-Central Agroforestry Research Institute, Jhansi

He is scientist, M.Sc. (Agroforestry) ICAR-Central Agroforestry Resear Institute, Jhansi 284003, Uttar Pradesh. His areas of interest are Forestry, Silviculture, Agroforestry with emphasis on stand management practices, Root Distribution and Structure, Policy and legal issues, Meta-analysis and Scientometrics.

Dr. Arvind Pai

Founder, Creative Insights, Bengaluru

Arvind Balakrishna Pai is a serial entrepreneur, a vocal proponent of designed in India, made in India products, he founded Creative Insights, (www.creativeinsights.in) a business consulting firm based out of Bengaluru in India, He has an experience of 3 decades in semiconductors and its applications, Indian design and manufacturing ecosystem and building new businesses. He advises hard tech and deep tech companies and start ups in identifying and building new businesses, new products and new revenue streams. Prior to this entrepreneurial start, he was working for Semiconductor companies like Philips Semiconductors, NXP and ST-Ericsson. Over a career of two decades, he contributed in product development, engineering, product marketing, business development and sales management roles. He built new clients, designed-in new products and generated over 100 M$ of sales revenues from India.

Dr. P.A. Pandya

Assistant Professor at Junagadh Agricultural University, Junagadh (Guj.), India

Dr. P.A. Pandya is presently serving as Assistant Professor at Junagadh Agricultural University, Junagadh (Guj.), India. He has worked/Working on Drip Irrigation, Agricultural and Meteorological Drought characterization and development of operation agricultural drought monitoring applications using remote sensing and machine learning, Hydrology, Watershed Management, Soil and Water Conservation. Teaching to UG (B.Tech, Agril. Engineering) and PG (M. Tech, Soil and Water Conservation/Irrigation and Drainage Engineering) at JAU, Junagadh, Gujarat, India.

Ms. Nisha Shah

Chief Executive Officer, NCCSD, Ahmedabad

Ms. Nisha Shah is working as a Chief Executive Officer, National Council for Climate Change, Sustainable Development and Public Leadership (NCCSD), Ahmedabad. NCCSD acts as an apex organization at the national level disseminating information and technology that will enable appropriate action and suggest policy frameworks to tackle impacts of global warming and climate change.

About the Book

Future Agriculture: Viksit Bharat

by

Kirit N. Shelat, Shrikant S. Kalamkar, A.R. Pathak and Odemari Mbuya

Indian agriculture has set new milestones in its progress. Since independence, major strides have been made in production of food grains, not only due to increase in area but also due to technology. As a result, the food grains production increased from 50.82 million tonnes in 1950–51 to 328.85 million tonnes in 2023–24[1]. The new commercial crops like sugarcane, cotton, jute, oilseeds also achieved a significant increase in its production relatively later than food grains[2]. The phenomenal growth in agricultural production since independence has been triggered by higher input use, particularly purchased inputs as well as technology induced productivity enhancement, massive extension efforts, improved farm practices and, above all, ingenuity and hard work of Indian farmers since the Green Revolution Period in late 1960s. However, several challenges[3] – some old and some new – remain. Growth of the agriculture sector has led to the unsustainable use of natural resources like land, water and bio diversity, spread of insects and pests, indiscriminate use of agro chemicals and adverse impact on ecology and environment. Despite noteworthy increase in per capita food production, some sections of the population still suffer from under nutrition and malnutrition. These challenges necessitate a paradigm shift in agriculture. In the meantime, new opportunities have arisen in the sphere of science and technology, information communication technology (ICT) and agri business which have the potential to transform agriculture production and post-harvest activities. There is a significant change on the demand side, with consumer preferences shifting towards healthy, safe, trait-based

[1] https://pib.gov.in/PressReleaseIframePage.aspx?PRID=2022740.

[2] Historically, the Green Revolution spurred a growth in the adoption of new technology and irrigation, which expanded the cultivation of wheat and rice at the cost of other crops like pulses, coarse cereals, and oilseeds (Joshi, 2019).

[3] Chand and Singh, 2023 (https://www.niti.gov.in/sites/default/files/2023-07/Aggricultrue_Amritkal.pdf)

and quality food and bios. These changes indicate that the future of agriculture (and those engaged within) will face profound transformation in the coming decades. There is a need to create an enabling environment for this transition, through appropriate policies and institutions, an enabling regulatory environment, development of frontier technologies, as well as public and private investments in agriculture and agri-business. This will enable agriculture to play a key role in achieving the goal of Viksit Bharat, inclusive development, green growth and gainful employment during *Amrut Kaal* (NITI Ayog, 2023). This book is an attempt to present the present situation, strategies to be adopted during the *Amrut Kaal* to make country *Atma Nirbhar Bharat* by 2047 having focus on agriculture and allied sectors. The final papers come out from the eminent experts are briefly summarized here.

Dr. Kirit Shelat, **Dr. Odemari Mbuya** and **Ms. Nisha Shah** in their article on 'Future Agriculture: Transformative Journey' noted the nation call given by the Hon'ble Prime Minister Shri Narendra Modi for collective action for rapid agriculture growth in *Amrut kaal* targeting doubling of farmers income, atmanirbhar bharat, sustainable development and eradication of poverty with suggested action points such as Natural Farming; Shree Anna; Skill – re-skill – up-skill – farmers, animal holders, fishermen; Smart use of water and harvesting of every drop; Smart use of Energy – Solar Energy – Bio – Energy; Micro-Enterprise at local level for processing by rural youth; Fisheries – Ocean area development; Climate/ Stress resilient farming, animal husbandry and poultry; and Exploring the new areas for income generations at village level. He mentioned that Government of India has already introduced a comprehensive package of schemes and initiatives which can enable this with proper implementation and its monitoring. He narrated the challenges of poor farmers, share croppers, and salinity ingress followed by ways to increase income. Further, he shared his experiences of *Krishi Mahotsav* – introduced by Hon'ble Prime Minister as the Chief Minister of Gujarat in the year 2004 which played important role in transformation of Gujarat agriculture.

Further paper highlighted that NCCSD initiated the Transfer of innovation-based technologies/practices for making agriculture climate resilient in view of Climate Change through promotion of quality seed of locally adaptable high yielding varieties of selected crops; informed management of natural resource base; demonstrations of proven innovation-based technologies for yield and income enhancements. It specially aimed at empowering farmers through skill development, up-skilling and re-skilling them to adapt to the experience harrowed below established that farmers are adopted technology add, To resilience, mitigations and add to income in the arena of climate change and enhances their income – double it despite adverse weather episodes. Authors mentioned that NCCSD has monitored the outcome with the help of Giotag report and undertook an impact assessment and details on each case study are presented.

Dr. V.K. Saraswat in his article on 'Water Security in Agriculture – Identifying Advanced Technology Confluence' noted that achieving water security in the agricultural sector is a multifaceted challenge that demands a holistic approach. Advanced technologies present a transformative opportunity to address water scarcity issues and enhance agricultural productivity, by embracing the prospects of AI and ML, IoT, digitalisation and circular economy principles, ensuring a more sustainable and water-secure agricultural landscape. As we navigate towards the Sustainable Development Goals (SDGs) for 2030, integration of these technologies into mainstream agricultural practices becomes not just an option but also a necessity for a water-secure and food-sufficient India.

While **Dr. Neelam Patel** in her article 'Empowering Women in Agriculture: Shifting Dynamics and Gender Balance' highlighted that rural women are the backbone of India's agricultural sector, yet they often face hurdles in reaching their full potential. We must empower these vital stakeholders to cultivate a more prosperous and equitable New India. By ensuring equal access to resources, technology, education, and healthcare, we can unlock the immense potential of rural women. Land ownership rights and skill development programs will further equip them to thrive. This holistic approach will not only boost agricultural productivity but also foster a generation of empowered and self-sufficient women, contributing significantly to a stronger nation. The government's initiatives in education, financial aid, and fostering collaboration among women are commendable steps in the right direction. These programs empower farm women, strengthening the agricultural sector and paving the way for a more secure and prosperous future for all.

Dr. K.D. Nayak & **Dr. Arvind Pai** in their article 'Technological Innovations for sustainable Agriculture' noted that India strides forward in its journey of progress and development, it is imperative to recognize the integral role of agriculture and the pivotal contributions of our farmers. From the historic achievements of the Green and White Revolutions to the present-day challenges and opportunities, agriculture remains central to our economic stability and food security. The confluence of traditional wisdom with modern technology holds the promise of a brighter future for Indian agriculture. By embracing innovations such as digitalization, precision farming, rainwater harvesting, robotics, artificial intelligence, and biotechnology, we can empower our farmers, enhance productivity, and foster sustainable practices. As we embark on this transformative journey, let us not forget the timeless values of resilience, perseverance, and unity, ensuring that every step we take towards agricultural prosperity echoes the spirit of Jai Jawan, Jai Kisan, Jai Vigyan, JaiAnusandan.

While **Dr. Meenesh Shah** in his article 'Sustainable Transformation of Dairying in India' stated that climate change has pronounced effect on feed production and nutrition of dairy animals. Increased environmental temperature would result in increased lignification of plant tissues which would affect feed digestibility and milk productivity. Water scarcity due to climate change would also affect feed and fodder production for animals. Climate change would increase incidence of vector-borne diseases, disease outbreaks, heat stress, reduce feed intake, milk yield and reproduction in animals. Climate change would also increase feed, water and shelter requirement of animals. If managed properly, agriculture and dairying would significantly contribute to mitigation of climate change and improve socio-economic sustainability. He suggested that to achieve the goal of Viksit Bharat, agriculture and allied sectors would have to focus on sustainable use of natural resources (land and water), mitigating greenhouse gas (GHG) emissions, increasing crop and animal productivity and efficiency in cost-effective manner, linking food production with health and nutrition (addressing nutrition through healthy diets), combating climate change, improving sustainability, modernization, digitalization, natural/organic farming, renewable energy use, and significant and sustained increase in farmers' income, etc.

Dr. V.P. Chovatia and **Dr. P.A. Pandya** in their article on 'Harnessing Secondary Agricultural Products for Economic Growth and Environmental Sustainability' noted that agricultural crops contribute significantly to the economy and environment of India, generating substantial waste annually from crop residues and byproducts. Mismanagement of this waste leads to air pollution, soil degradation, and inefficient resource utilization, causing billions of dollars in economic impact due to lost revenue and increased healthcare costs. To address this challenge, initiatives focusing on waste-to-energy technologies, efficient recycling methods and improved farming practices are being explored. Furthermore, the utilization of secondary agricultural products is vital for maximizing economic potential, especially for smallholder farmers. Despite challenges like limited infrastructure and technological know-how, investments in research, capacity building, and policy support are essential. Harnessing secondary agricultural products is critical for increasing farmer income, promoting sustainable development, and mitigating environmental impacts. The importance of secondary agricultural products extends beyond economic benefits to climate resilience and sustainable agriculture. These products contribute to bioenergy generation, carbon sequestration, income diversification, food loss reduction, and livestock feed improvement. Leveraging these products can enhance climate-smart practices, adaptive capacity and overall agricultural sustainability, biocides enhancing farmers income.

Dr. C.K. Timbadia in his article on 'Enhancing Soil Carbon Sequestration Through Natural Farming Practices: A Sustainable Solution for Climate Change Mitigation and Food Security' highlighted that growing global population's demand for food is compelling the agricultural sector to adopt advanced technologies, replacing traditional practices. Consequently, the sustainability of crop production systems, which relies on soil quality was being impacted by the farming methods to be employed, such as, Intensive crop cultivation using imbalanced fertilizer, high nutrient mining through monoculture, excessive tillage and inversion coupled with the removal of crop residues by burning, hastens the decomposition of soil organic matter. This process can result in significant soil carbon loss. He mentioned that Natural Farming is increasingly advocated as an alternative approach to combat soil degradation caused by conventional agricultural practices that deplete soil fertility, with the aim of achieving higher crop productivity as a short-term benefit. Natural Farming practices are recognized to improve SOC contents in Natural Farming involves fundamental principles such as minimal soil disturbance, maintaining permanent soil cover or using cover crops, practicing mixed cropping, mulching with crop residues, utilizing on-farm bio-inputs such as inoculum cultures of beneficial microbes and fermented botanicals for pest control. Therefore, addressing the dual challenges of food insecurity and climate change can be achieved by restoring soil carbon through the adoption of Natural Farming practices.

Dr. Arunachalam, **Dr. S. Suresh Ramanan** and **Dr. A.K. Handa** in their article on 'Agroforestry in the context of climate change for future agriculture' note that agroforestry is a sustainable land use system for food production and livelihood sustenance in the ongoing era of warming temperature due to climate change. This tree-based farming system has huge potential to mitigate and adapt the growing impact of changing climate. Additionally, agroforestry can address other multiple challenge like food security, biodiversity conservation, and sustainable management of natural resources. The integration of trees with crops plays a sustainable role in achieving several sustainable development goals (SDGs) including SDG-1 (No Poverty), SDG-2 (Zero Hunger), SDG-12 (Sustainable Consumption and Production), SDG-13 (Climate Action), and SDG-15 (Life on Land). Agroforestry is the need of hour for future food production in context of changing climate in developing countries like India. Although agroforestry in various forms like home gardening has been an age old practice in India, the institutionalization of agroforestry research in India began in 1979 with National Agroforestry Seminar at Imphal organized by Indian Council of Agricultural Research (ICAR). In this journey of about four decades, several milestones were established like formulation of All India Coordinated Research Project on Agroforestry (1983), foundation of Central Agroforestry Research Institute at Jhansi (1988), and launch of Agroforestry Policy (2014), Submission on Agroforestry (SMAF, 2016–17), and Accreditation Protocol for Agroforestry Nursery (2024). However, we have to go long to make agroforestry the people's practice and strengthening human well-being in India

Dr. A.R. Pathak, **Dr. P.A. Pandya** and **Dr. D.K. Varu** in their paper on 'Prospects of Digital Agriculture in Gujarat' examine the application of digital technology across various aspects of agriculture, illustrating its transformative potential and key considerations for adoption. They highlighted that with the advent of digital technologies, the agricultural landscape in Gujarat is undergoing a profound transformation. Digital agriculture encompasses the use of various technologies such as precision farming, IoT (Internet of Things), remote sensing and data analytics to optimize crop production, minimize resource wastage and enhance farm efficiency. With the world population growing and resources becoming scarcer, the need for sustainable and efficient agricultural practices is paramount. Digital technology offers a suite of tools and solutions to address these challenges, enabling farmers to optimize resource use, minimize environmental impact, and enhance crop quality and yield. They narrated the challenges in adoption of digital agriculture in Gujarat. One of the primary challenges is the digital divide, with many smallholder farmers lacking access to technology and digital literacy. Limited internet connectivity and electricity supply in rural areas further exacerbate this divide. Additionally, high initial investment costs and the complexity of implementing digital solutions pose barriers to adoption for some farmers. Addressing these challenges requires concerted efforts from government, private sector and civil society organizations. Looking ahead, the future of digital agriculture in Gujarat appears promising. Advances in technology, such as AI (Artificial Intelligence) and machine learning, hold immense potential for further optimizing agricultural practices and decision-making. Moreover, initiatives aimed at bridging the digital divide, such as rural internet connectivity projects, will help ensure that all farmers can benefit from digitalization. By embracing digital agriculture, Gujarat can not only enhance its agricultural productivity and sustainability but also empower farmers and strengthen rural economy.

Dr. Rajendra Shende in his article 'Smart, Sustainable and Space-Agriculture' highlighted to leverage Akash (space) through Space Seed Breeding Programme to enhance the productivity when those seeds are brought back on the Erath. University campuses are the breeding grounds for such innovations.

Dr. V.V. Sadamate in his article 'Transformative Agricultural Extension Strategies for Future: Policy Implications and a Way Forward' mentioned thatnearly 83 per cent of farmers are small and marginal ones and thus extension delivery is a huge task to cover over 700 districts, around 7000 blocks and nearly 7 lakh villages covering 127 agro-climatic zones and 14 Crore Farmers is a huge task. The farming population is spread over varied social dimensions, varied geographical situations, varied resource conditions. Therefore, agricultural extension is a tough articulation and architecture to make a significant impact. He suggested for urgent extension innovations/reforms for future. Further advocated/suggested for delivery approaches, strategies and models that needs to be get transformed to deal with the future outreach challenges in agricultural and allied sectors.

Dr. Suresh Acharya in his paper on 'Unlocking Potential: Sustainable Development in Arid Areas' highlighted that India's arid zones span approximately 35 million hectares across seven states, with Rajasthan and Gujarat contributing over 80% of this area. Despite facing numerous challenges, these regions also offer significant potential for sustainable development. Further he noted that the primary challenge in arid areas lies in the scarcity of freshwater and advocated that a multifaceted approach is pivotal for managing challenges in arid regions. Water management is critical, and strategies such as rainwater harvesting, desalination (considering costs and brine disposal), smart irrigation, and wastewater treatment are essential. Empowering local communities through training and shared decision-making, while integrating indigenous knowledge, is vital for sustainable development. Sustainable agriculture, responsible tourism and promoting local crafts can create economic opportunities and improve livelihoods. Renewable energy sources, such as solar and wind power, can reduce dependence on fossil fuels and contribute to a greener future. Combating desertification and preserving biodiversity are also crucial for the long-term environmental health of arid regions.

Dr. A.R. Pathak in his paper on 'Shree Ann—Opportunity in Gujarat' mentioned about year 2023 as the International Year of Millets declared by The Food and Agriculture Organisation of the United and importance of millets in context of India and Gujarat. He noted that Millet crops are drought tolerance require less irrigation can be grown in light soil and climate resilient can withstand against abnormal weather conditions. Being a C4 plants more carbon absorption and sequestration hence improve soil health. From nutritional point of view, contains fibre, protein, vitamins minerals, antioxidants and Phyto chemicals higher than other cereals. He noted that in Gujarat, awareness of Natural Farming is increasing and area is also increasing, there is great opportunity to grow millets under Natural Farming, as these crops require less fertilizer and tolerant to pest and diseases with less irrigation. Due to Natural Farmers will get better Produce available to consumers without chemicals.

Dr. Odemari Mbuya and **Dr. Kirit N. Shelat** in their article Agriculture and Carbon Sequestration: A Climate Change Mitigation Strategy' highlighted the issue of excessive use of fossil fuels, changes in land use and land cover patterns during Green Revolution and after phase, which have inadvertently resulted into a sharp rise in concentration of greenhouse gases in the atmosphere. Further, they noted that agriculture serves as both a source and sink for GHGs and suggested to effectively reduce the current amount of CO_2 in the atmosphere, we must increase the carbon sink area and increase the carbon sequestration efficiency.

Dr. Gopichandran in his article presented public policy outlook on future agriculture at the interface of climate change and chemical ecological and ecosystems level perspectives of bioresources management. Author argues that one of the central features of public policy is the imperative of science at its core due to evidences it provides. This extends seamlessly into technology applications with special reference to benefits for the marginalized in particular. With respect to the former it is essential to explore domains that fine grain our understanding of resilience pathways, across the soil, water, air media as carriers of chemicals that mediate interactions and directly so on bioresources. The IPCC has periodically highlighted facets of climate resilient development pathways. Our understanding of planetary boundaries too has evolved; only to reinforce the call for rapid, yet sound mitigation and adaptation measures. Climate resilient agriculture is a case in point. It is essential to base related policies on principles of ecosystem level resilience, that are in turn driven by stochastic and non-stochastic processes and thresholds. System specific insights are therefore critical to assess individual, synergistic and antagonistic influences, duly respecting the time – scales over which they manifest. The present narrative is therefore a call to take note of emerging facets of ecology, especially of allelochemical and allelopathic pathways that could determine forms and functions of sustainable crop systems. The other aspect is about the impact of high levels of UV B exposure of crops due to depleting ozone layer.

Dr. Sanjay Deshmukh in his article on 'Agroforestry for the Future: Driving Sustainability, Securing Livelihoods, and Combating Climate Change' highlighted that the importance of agroforestry extends beyond its immediate benefits to farmers. It is critical in mitigating climate change through carbon seq uestration, reducing soil erosion, and improving water retention and quality. By enhancing biodiversity, agroforestry systems contribute to the resilience of agricultural landscapes, making them more capable of withstanding and recovering from environmental stresses. This integrative approach also offers economic advantages by diversifying income sources for farmers and creating employment opportunities in rural areas. Author conclude that the transformative potential of agroforestry lies in its ability to create win-win scenarios for people and the planet. It offers a sustainable way to meet a growing population's food, fiber, and fuel needs while conserving natural resources and mitigating climate change. Integrating trees into agricultural systems can enhance productivity, build resilience, and create sustainable livelihoods. The journey towards sustainable agricultural futures through agroforestry is not just a possibility but a necessity.

Dr. Odemari Mbuya in his article on 'Soil Health' highlighted that Soil health is the continued capacity of soil to function as a vital living ecosystem that sustains plants, animals, and humans, and connects agricultural and soil science to policy, stakeholder needs and sustainable supply chain management. Soil health is essential for agriculture, and crucial to many other ecosystem services. Quantifying soil health is still dominated by chemical indicators, despite growing appreciation of the importance of soil biodiversity, due to limited functional knowledge and lack of effective methods. He opined that achieving healthy soil with the right balance of minerals, organic matter, water, air, and microorganisms isn't easy, that's why you need help from professionals. In intensive modern agriculture, the role of science, engineering and technology in maintaining a healthy soil can neither be overlooked nor ignored.

Contents

Future Agriculture: Transformative Journey

Kirit Shelat[1], Odemari Mbuya[2] and Nisha Shah[3]

Introduction

Hon'ble Prime Minister Shri Narendra Modi has called for collective action for rapid agriculture growth in *Amrutkaal*:

- Double Income
- Atmanirbhar
- Sustainable Development
- Eradicate Poverty

He has suggested many action points – important include:

- Natural Farming
- Shree Anna
- Skill – re-skill – up-skill – farmers, animal holders, fishermen
- Smart use of water and harvesting of every drop
- Smart use of Energy – Solar Energy – Bio-Energy
- Micro-Enterprise at local level for processing by rural youth
- Fisheries – Ocean area development
- Climate/Stress resilient farming, animal husbandry and poultry
- Exploring the new areas for income generations at village level

India is on the way to *Atmanirbharata* and the Developed Nation *Amrutkaal*. Key for rapid development and income increase in the agriculture sector is agriculture with integrated links with water and energy and their smart use for food and agricultural systems.

[1] Former Principal Secretary, Govt. of Gujarat & Executive Chairman, NCCSD, Ahmedabad, India.

[2] Professor of Agricultural Sciences, Director of the Center for Water Resources at Florida Agricultural and Mechanical University (FAMU).

[3] Chief Executive Officer, National Council for Climate Change, Sustainable Development and Public Leadership (NCCSD), Ahmedabad, India.

Goals for Amrutkaal

- Win the challenges of climate change and its adversity.
- Continue doubling income of farmers every five years – despite adverse weather events.
- Remove poverty.
- Increase production and productivity and growth.
- Bring up productivity level.
- Food and nutritional security.
- Generate local and international markets and make available higher returns to farmers. Make Agriculture profitable.
- Generate rural wealth by local level micro-enterprise and processing.
- Contain inflation and meet consumer demand.
- Provide multiple sources of income like honey production, seaweed.
- Input links – identify fraudulent producers and punish them.
- Remove income disparity between rural and urban families.
- Smart use of water and energy and developing integrated nexus of food, water and energy with focus on increasing income in rural areas and that of farmers.

Current Situation and Trends

- Agriculture growth is stable and sustainable.
- Massive Horticulture production and expansion in Animal Husbandry.
- Outstanding work in Natural Farming.
- Country's agricultural administration is well prepared to achieve all goals and has the capacity to move farmers to produce more with less.
- Average growth has remained around 3 to 4 per cent per annum in the last decade.
- Country is self-reliant for its food security and now a major exporter.

I. *Current Trends: Challenges*

1. Adverse intense weather events. Cyclone, Flood, Heavy single day rain episodes, delay in intervals, unseasonal rains, heat and cold wave, dust storm, locust attack – news pests and insects and overall increase in temperature.
2. Some adverse impacts are irreversible like melting of glaciers in the Himalaya – similar rise in sea level globally shrinking ice cover over Greenland and Antarctica is another irreversible change. All these

have serious impacts on coastal areas and Himalayan region – which need attention.

3. Farming is becoming risky. Farmers like to give it up. Tremendously increasing non-agricultural activities – reducing land and escalating new share-croppers – '*Bhagias*'.
4. Our per capita emissions are too low compared to the world average. But we are the third largest emitter of carbon dioxide (CO_2) and that has continuing adverse impacts.
5. Increased demand/agitation for irrigation water and energy due to increased temperature.
6. Younger generation does not like the management of cattle like milking, collecting cow-dung, taking out cattle for grazing or work in farms.
7. Deterioration of soil health and water quality.
8. Poor market linkage.
9. Rise in expectation of farmers for higher income and greater return for their produce.
10. Increasing need of funds for providing relief to farmers due to adverse weather events and for procurement under Minimum Support Price (MSP).
11. Ponds – in villages getting dried up – and even with all the infrastructure we have – water needs to be supplied by tankers in quite a few villages and villagers have to walk long to fetch it.

Challenges – Poor Farmers

There are four types of farmers:

- Rich and/or progressive/educated farmers
- Poor farmers
- Sharecroppers – *Bhagias*
- Women farmers.

Two-three types of Animal Holders:

- Farmers' Owners of milk cattle connected with the Dairy network.
- Shepherds – owners of other animals like sheep – goat – pig.
- Poultry farmers.

Poor farmers, Sharecroppers, and small animal holders – and women farmers need attention.

ICAR needs to develop different guidelines/advisory as per need and capacity of understanding of each group keeping in view – Block Level Agro – Climatic Condition.

Challenges – Salinity Ingress

- Due to rise in sea-level and faults -underground and over ground – in ocean areas – salinity ingress is causing havoc with water resources – which are becoming contaminated. Crop productivity nearer to ocean areas is severely affected.
- As per Central Pollution Control Board – using date of National Water Quality Monitoring Program 28 districts of Gujarat have high salinity level, high fluoride level, nitrates, arsenic and iron and even excessive lead in some districts. But due to Narmada Dam water situation is safe in Gujarat.

Remove Poverty

- Current system is doing extremely well – but in the same village with similar land resources, one farmer makes profit, the other legs behind and remains poor.
- Poor farmers and shepherds need different strategies and door- step contact at village level. They do not come to the block level and are left out of the development process.
- To remove poverty – first poor farmers/animal holders/fishermen need to be identified and given focused attention and monitoring.
- Deliver them affordable technology and multiple sources of income and support to their young.
- Identify poor farmers village-wise and assist them individually and monitor their progress. The Rural Development Ministry can undertake this with Panchayati Raj Institutions at ground level.

Sharecroppers

Increase in the number of sharecroppers and land under that is a worrying trend. Sharecroppers do not get any bank loan nor entitled to any benefits of Government Schemes due to lack of land ownership.

Sharecroppers: include

- Remote area farmers who migrate to work in urban centers
- Rich farmers – who have shifted to urban areas for education of their children and/occupation.

Lands by developers – for housing and commercial development. Such areas are in the periphery of all urban centers and metropolitan cities have such lands up to 100 km radius.

This needs to be addressed. This is the one reason for stagnancy in productivity.

II. *Ways to Increase Income*

New Cultivable Areas

- Country has seen changes in weather patterns in the last two decades. Earlier drought prone areas have turned green – including desert areas which are receiving good rains in states like Rajasthan and Gujarat.
- This has brought new cultivable areas on the margin of desert – both great and little Ranns of Kutch and in Rajasthan near Jaisalmer and other places. There are existing arable wastelands also.
- This is opening up opportunities for a huge livestock-based economy. Such lands can be used to produce grasses, fodder trees and salinity resistant crops such as date palm, Subabul, Acacia, Arabica – to support a massive livestock economy. Drinking water can be provided by solar desalination plants. Such an economy will provide huge employment, income and CNG (Compressed Natural gas) gas and cow-dung which can be used to restore fertility of lands. There is growing demand for dairy and milk Products.

Reduce Losses to Farmers

- ***Transit loss:*** Farm to market point. Estimated losses are about 25 per cent to 40 per cent.
- ***Distress Sale:*** Poor farmers are not in a position to take the produce to the MSP Centre. Hence local sale at less price. MSP centers can collect produce from village level like Milk Producers' Union do for collection of milk. This can be done in a pre-announced manner.
- ***Spurious Inputs:*** As per one study of Department of Consumers Affairs – more than 50 per cent seeds are spurious and similarly fake mixed fertilizers and pesticides are sold. Farmers lose productivity. There is a need to severely penalize fake producers and distributors. There is also a need to make farmers aware of spurious products.

Seaweed

- Country has a huge coastal line. States like Gujarat have 1,600 km.
- While rise in sea level is causing concern, Sea provides vegetation like seaweed. Seaweed grows in sea water in coastal areas. It is a floating crop. It has multiple uses. In some countries it is used as food. But it

has industrial value. It is used as a bio manure, while its extract is used as bio pesticide. It also absorbs CO_2 and most importantly flourishes in saline Seawater. It can provide an additional source of income to farmers/fishermen. It can be cultivated on seashore by setting up ponds similar to what we have for salt farms.

Generate Wealth at Village Level

- Key is local processing
- Technology is available for mini-plants. Some are already under use.
- This will create micro enterprises and one man/two men units – value addition and income will generate at village level.
- Develop Bankable Schemes. Apart from Agro-processing – following can be included.
    - Services to Urban Centres – like gardening, cleaning, providing tiffin services in industrial units etc.
    - Tool bank – includes Tractors – Rotavators, Drones etc.
- Many villages now have three phase power – round the clock electricity supply. This can support local processing which will generate the wealth at village level and reduce transportation cost and wastage and increase income of farmers and local employment.

Energy Security – Solar Energy

- Solar Schemes and technology are available but not used by the majority of rural households and farmers – due to lack of knowledge or availability at local level. There is no functional unit of Renewable Energy Department at block level to guide – for getting it, using it, and maintaining it including applying for schemes and sale of surplus energy generated to electricity companies.
- Taluka and Gram panchayats need to be made responsible for the task of promoting roof solar energy for every house and in farms.
- Sale of excess production to local power supply companies. This is already introduced but with limited farmers.
- Solar energy reduces the energy cost and its availability during the day time is convenient for farm operation – it will provide an assured source of income by its sale.

Smart Villages

- *Smart Villages:* There is a scheme called *Shyama Prasad Mukherjee RURBAN Mission* launched by the Government in 2016. It is important that maximum advantage of the scheme is taken since the majority of villages are well connected by rail & rural roads, electricity & LPG, mobiles and WiFi, and TV channels. Urbanites can comfortably live in such villages.

- Now the question is how to use all these effectively to create production and services base in rural areas and small towns and meet aspirations of local youth for employment and prevent migration.

Bioenergy

- *Bio Energy:* There is considerable agricultural waste. Farmers partly use the same to mix with cow-dung for composting. Surplus is wasted. Some burn it. Gujarat has biofuel making brick units. There are about 200 of them. They provide cheaper fuel to industry and income to farmers. Such units can be set up, one for every village. This will solve the problem of burning of crop residues in farms and increase income of farmers.
- *CNG Gas:* Banas Dairy – Gujarat has set up a CNG unit based on Biogas – produced by cow-dung. This is a successful venture. It provides additional income to farmers who get slurry and payment for cow-dung. This will increase the income of farmers. This can be replicated by all Dairies.
- Bio-CNG contains 92–98 per cent methane. The calorific value is 13000 kcal kg. Banas Dairy – Gujarat has set up a 3500 cubic meter biogas reactor. Plant uses cattle dung from 250 farmers along with milk collection. Bio gas is compressed and stored in cascades. Cascade delivered to Bio-CNG filling station.

Increase in Income-Carbon Credit

- Country has massive horticulture growth – fruits – vegetables – spices.
- Fruit trees are entitled to carbon credit.
- The Ministry of Environment has set up a separate cell with funds for "Green Credit" in which sustainable agriculture is included.
- There is a need to promote farmers and their organizations for obtaining carbon credit. This will provide an additional source of income to farmers irrespective of adverse impact on productivity due to weather events.

New Dimension for International Market

- Currently the Ministry of Commerce has Commodity Boards. They are responsible for production, development and exports. These boards are for Tea, Coffee, Rubber, Spices and Tobacco. These products have already stabilized. They do not seem to need any Future support.

- But country has opportunity for meeting international demand for many other products like i) Horticulture-fresh Fruit and vegetables, ii) Fisheries, iii) Castor oil and derivatives, iv) Organic products, v) Dairy products, vi) Food grain Processed food products & ready to eat food products.
- The Ministry of Commerce needs to paradigm shift and reorganize, wind up old boards or re-designate as per current needs to boost the market of crops and produce and products which have demand.

ICAR – Restructuring

- ICAR is India's premier scientific research, education and development Institute. It has the largest pool of agricultural Scientists in the world.
- ICAR has transformed Indian Agriculture and brought about sustainable growth.
- In the changing world and challenges, ICAR has to be strengthened and deal with priority areas, e.g.
    - *Climate Change:* ICAR with its network of KVKs (Krishi Vigyan Kendras) and State level Agriculture Universities along with network of Research centers provide farmers at Village local weather advisory followed by Agro Advisory, round the year. This is the key to sustainable agriculture in the arena of Climate Change.
- Similarly Research Centres have to be integrated with cultivation to end use and Market analysis and intelligence. And make available advice to farmers at Village level using digital technology. The Research Centre – which has very good land resources – must become Atmanirbhar and showcase the use of all technology and smart practice which is being advocated to farmers. They have to be the center of smart use of food water energy nexus. ICAR has to broaden its vision and expand work on water and energy use.
- It is equally important for ICAR and its affiliated universities and Research Centres to put in public domain work done.
- Major criticism by farmers is, while they are given advice, the same is not followed by such Research Centres who have huge land resources like cultivation based on current soil health and water analysis, use of moisture meter, drip and sprinklers, and so on and so forth.

Water Security

Water is fundamental to all living beings. Water is a key to sustainable agriculture. Water is also key for continuous increase in income of farmers. Farmers must have the opportunity to take more than one crop. Assured irrigation is therefore needed.

(a) *Harvested Water: Irrigation Department*

- Efficient use by reducing wastage while transferring to farm.
- There are leakages, theft and inefficient discharge. Losses, as per one estimate, up to 50–60 per cent. But there is no accounting of (a) potential vis-à-vis actual discharge, (b) what is discharged and what is actually delivered to farmers. Hence like in the energy sector, distribution should be separated from storage. Account for every liter of water which is discharged and measure wastage. Strict vigil to reduce all wastage during transit and reaching out to all villages to use full potential and set up a system of accountability for efficient use of total potential.
- Maintenance of canals, check dams, ponds is a problem. Very many harvesting structures are silted.
- Panchayat Raj institutions should be given responsibility for it. They need to maintain all canals, check dams, ponds, many of which are silted. They need to connect all farms with piped water supply.
- Finally, farmers, if they want irrigation, must use drip systems. No more flow irrigation.
- Evaporation losses are high. They are increasing with increasing temperature.
- All canals can be covered with Solar panels. Contracts can be given to private players and income can be generated. Ponds both in villages and urban centers can be covered by solar panels.

(b) *New Steps for Water Conservation*

Introducing integrated River Basin Management in all rivers and revaluates. Compulsory recharge and recycling of wastewater by all civic bodies both urban and rural. Compulsory recharge by all housing and infrastructure projects both existing and future, urban and rural.

(c) *Prevent Salinity Ingress*

This is causing water contamination. Construct prevention walls, dams or dykes. Create natural barriers by planting Bamboos.

(d) *Make Stakeholders Responsible*

- Farmers must use drip, have drainage/trenches/farm ponds in the farm. Use soil moisture meters to determine the need of soil for irrigation. Every house holder existing or future needs to have recharged wells.
- Civic bodies recycle and recharge. Infrastructure projects like Roads, Railways must be done with drainage systems and connected with recharge wells. Irrigation Department to be made accessible. This will need strong monitoring.

Flood Prone Area Scheme

- There are increasing events of one day heavy rains, accumulated heavy rain, for two to three days cyclones. These events wash out fertile soil and damage livelihood, assets.
- Flood prone areas need Action Plan.
- Integrated River, Ponds management Schemes.
- Drainage in all infrastructure projects with re-charge wells-roads-railway-ports.
- Compulsory recharge wells by all civil authorities – Panchayats – urban bodies and all housing existing and future.
- Restoration of washed-out Agricultural lands by bringing silt from where it has got accumulated.

Technology

There are Three Areas

Existing Technologies: Already Popular – Tractor, Pump Rotavators, Harvester, in many cases they have higher capacity then needed.

- Technologies – useful but not adopted by all farmers – like drip, green House, soil health and water analysis.
- New Technologies: AI, GIS, Remote Sensing, precision farming,
- Three types of farmers:
    - Progressive and or Rich
    - Poor
    - Sharecropper
- There is a need to study what is convenient to actual user and what is actually needed.
- ICAR can work on this suggestion and guide affordable technologies and guidance as per the category of farmers.

Agriculture mainly depends on the nature. However changing climate and global warming are making farming unpredictable. The need to use modern technologies to increase productivity and profitability has therefore led to the adoption of Agriculture 4.0.

There have been significant changes in the context of agriculture over the decades with development of many new technologies. Several new-age farmers are now using soil mapping software to determine the optimum level of fertilizers used in the farms.

Application of these emerging technologies in farming and agriculture has paved the way for more opportunities. The Agro-tech Start-ups and traditional farmers are now said to be using the latest solutions and trends to improve production in the food value chain, including the adoption of new technologies.

NCCSD initiated the Transfer of innovation-based technologies/practices for making agriculture climate resilient in view of Climate Change through promotion of quality seed of locally adaptable high yielding varieties of selected crops; informed management of natural resource base; demonstrations of proven innovation-based technologies for yield and income enhancements. It specially aimed at empowering farmers through skill development, up-skilling and re-skilling them to adapt to the experience hirroted below established that farmers are adopted technology add, To resilience, mitigations and add to income in the arena of climate change and enhances their income – double it despite adverse weather episodes.

To promote sustainable agriculture in arena of climate change to build a climate smart farmer by

- Introduction of climate resilient practices and technology
- Smart use of Soil, Water & Energy
- Promote mitigation to reduce GHG
- Increase farm productivity and income
- Increase use of natural farm practices
- Introduction of Weather advisory followed by Agro Advisory
- Kachchh (Ta Nakhatrana); Amreli (Ta Rajula) and Bharuch (Ta Jambusar)

Technology Transfer at Farmer's Field

1. Improving Soil Health Sustainable Inputs (1 acre) (Cotton, Mango, Dragonfruit & Date palm and other crop) 60 Model Farm at Rajula, Jambusar and Nakhtrana)	**2. FYM Enrichment** 63 (21 Model Farm Rajula + 21 Jambusar + 21 Nakhtrana)	**3. Water Management & Energy Consumption** Moisture Meter 45 (15 demonstration at Rajula + 15 Jambusar + 15 Nakhtrana)
4. Laser Irrigation System (1 acre) 15 (5 Demonstration at Rajula + 5 Jambusar + 5 Nakhtrana)	**5. Pest Management** Pheromone Trap (1 acre) 63 (21 demonstration at Rajula + 21 Jambusar + 21 Nakhtrana)	**6. Smart Use of Land** Saves Water & Fertilizers Compare to Conventional method Trellis system in Vegetable (0.25 acre) 5 (2 demonstration at Rajula + 2 Jambusar + 1 Nakhtrana)
7. HDP plantation – Horticulture crop (1 acre) 3 (1 Rajula + 1 Jambusar + 1 Nakhtrana)		

NCCSD has monitored the outcome with the help of Giotag report and make an impact Assessment. Detail case studies are as under.

Case Study-1: Improving Soil Health by Using Sustainable Inputs.

Sustainable Inputs

In sustainable agriculture, the goal is to reduce the input of external energy and to substitute non-renewable energy sources with renewable sources. Sustainable agriculture, also known as sustainable farming, is defined as producing food and livestock over the long term with minimal negative effects on the environment. Sustainable inputs ensure that crops get the required nutrients and produce better quality products at lower costs.

Following Sustainable inputs are given to the 60 farmers of the three districts.

Bio-enriched organic manure

Mahalaabh
Herbovita
13-00-45
Micromix

Boricha Devatbhai Ravatbhai from Amreli Districts (Latitude–20.953450 & Longitude–71.453448) is using sustainable inputs at their farm. By using Sustainable inputs, he found that Excessive use of chemical fertilizer is been decreased, because Sustainable inputs reduces the biological nitrogen fixation in the soil and increases the chemical pollutants of nitrous oxide.

Sustainable agriculture integrates three main goals—environmental health, economic profitability, and social and economic equity; with that goal 60 farmers from the three districts (Amreli, Bharuch & Kutch) are using the sustainable inputs at their farm.

Case Study-2: Improving Soil Health by Using FYM Enrichments.

FYM Enrichment

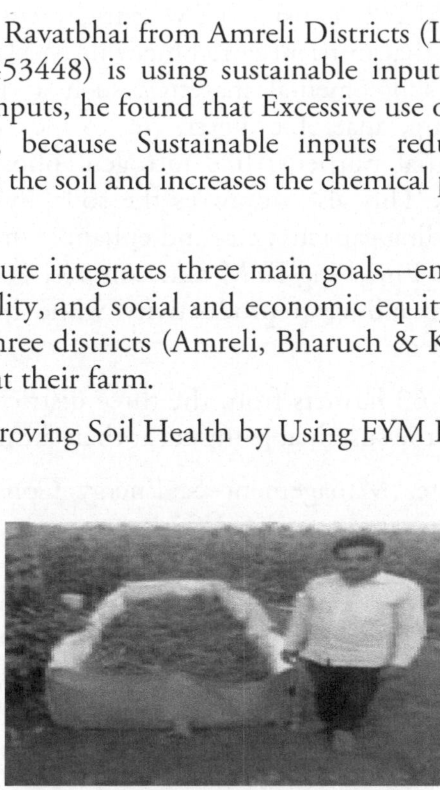

Farmyard manure refers to the decomposed mixture of dung and urine of farm animals along with litter and left over material from roughages or fodder fed to the cattle. Enrichment of FYM and compost means improving the quality of these organic manures by adding any specific substance that can increase the nutrient content and microbial population FYM and compost are widely used in organic farming as they supply plant nutrients and improve soil properties.

Following Sustainable inputs are given to the 63 farmers of the three districts.

Vermibed (1 no)
Compost Culture
PSB Culture
Metarhizium
Trichoderma
Jeggary

Sojitra Jitubhai Raghavbhai from Amreli Districts (Latitude –21.093663 & Longitude–71.451272) is using FYM Enrichments at their farm. By using this he found that nutrient enrichment in FYM and compost is the process of adding supplemental materials such as bio fertilizers, rock phosphate, Woodchip, anaerobic digest, etc. to increase the content and availability of essential nutrients like nitrogen, phosphorus, potassium, and micronutrients. This also improves the soil physical properties like structure, water holding capacity, etc. and enhances the microbial Activity and decomposition. Enriching FYM and compost is an effective way to improve soil health and crop productivity while minimizing negative environmental impacts.

To enrich their soil 63 farmers from the three districts (Amreli, Bharuch & Kutch) are using the sustainable inputs at their farm.

Case Study-3: Water Management & Energy Consumption by using Moisture Meter.

Moisture Meter (For scheduling irrigation)

Sufficient water saturation is vital for plant development and it is the farmer's foremost task to maintain it. Moisture meters are used to measure the amount of water (Moisture) is present in the field. Measuring the moisture content of our soil or compost pile allows us to analyze whether we need to add more or less water. Water is so important for carrying nutrients into plants and facilitating the composting process. Too much or too little water can cause great problems which are why measuring with these meters can allow us to get water content just right. Moisture meter are given to 45 farmers of three Districts.

Jadav Gitaben Ramsangbhai from Bharuch Districts (Latitude – 22.157299 & Longitude – 72.838439) is using Moisture meter in her farm. While using this meter she noticed such benefits like

- It saves water up to 5–8 lacs lit per hectare
- Require less water quantity
- Reduce water borne disease
- Better root growth
- Improve nutrient status by reducing soil salinity

To 45 farmers of three districts (Amreli, Bharuch & Kutch) are using this moisture meter in their farm.

Case Study-4: Water Management & Energy Consumption by using laser Irrigation.

Laser Irrigation System

Laser irrigation is an innovative alternative to drip and sprinkler irrigation techniques punched with laser holes at definite intervals to discharge minute droplets to the crop with both laser spray and laser drip irrigation.

Laser irrigation can be adopted in a wide range of crops from suitable for leafy vegetables, onion particularly they enhance the humidity and alter the micro-climate for better yields in the summers particularly. Micro Irrigation like Drip irrigation, Sprinkler irrigation, Rain gun Sprinkler irrigation are in use but expensive for small Farmers Laser irrigation is cheaper and efficient system.

Laser Irrigation System is given to 15 farmers of three Districts.

Parmar Ranjanben Sanjaybhai from Bharuch Districts (Latitude – 22.168810& Longitude - 72.87499) is using Laser Irrigation System in her farm while using this System she noticed such benefits like:

- Utilized 90–100% cultivable land
- Water saving up to 40%.
- Reduction in Energy consumption.
- Prevent soil erosion
- Increase in productivity up to 30%

It is affordable, less maintenance and wets field like rain. It is also applicable in hilly regions. It saves water, Energy, helps in controlling pests, diseases and improves soil health. 15 farmers from three districts (Amreli, Bharuch & Kutch) are using this system in their farm.

Case Study-5: Pest Management by using Pheromone Trap.

Pheromone Trap

A pheromone trap is a type of insect trap that uses pheromones to lure insects. Sex pheromones and aggregating pheromones are the most common types used. Pheromones are chemicals used by insects and other animals to communicate with each other. Insects send these chemical signals to help attract mates, warn others of predators, or find food. Using specific pheromones, traps can be used to monitor target pests in agriculture. Pheromones are chemicals used by insects and other animals to communicate with each other. Insects send these chemical signals to help attract mates, warn others of predators, or find food.

Pheromone Traps given to the 63 farmers of the three districts.

Jadav Jamnaben Dalpatbhai from Bharuch Districts (Latitude – 22.161826 & Longitude – 72.836962) is using Pheromone Trap in her farm, while using this System she noticed following benefits.

- Affordable, easy to install and manage
- If used properly can detect low numbers of insects
- Collect only species of interest
- Non toxic
- Can be used all season long
- Can be used to monitor for specific exotic pests.

Using specific pheromones, traps can be used to monitor target pests in agriculture and for that this trap are using by 63 farmers of the three districts (Amreli, Bharuch & Kutch) in their farm.

Case Study-6: Smart Use of Land—Saves Water & Fertilizers Compare to Conventional method by using Trellis System.

Trellis System in Vegetable Crops

A trellis is a type of structure used to guide and support climbing plants such as ivy. The purpose of the trellis is to support climbing plants, which can't stand straight like a tree or houseplant. With the aid of the trellis, the climbing plant wraps around the trellis. A trellis may be constructed from various materials, but the common denominator is an open structure that facilitates vertical plant growth. Trellis System is given to the 5 farmers of the three districts.

Dakhada Amrubhai Vajsurbhai From Amreli District (Latitude – 20.949667 &Longitude-71.450613) is using Trellis System in his farm, while using this System he noticed plants' exposure to sunlight is increased and this makes the plant to grow and yield well too. Generally, improve the quality of fruit or foliage – The plant does not experience rot before harvest. This system promotes healthy crops and with this Pruning and fertilizer application becomes easier.

Trellis System is given to the 5 farmers of the three districts (Amreli, Bharuch & Kutch) in their farm.

Case Study-7: Smart Use of Land—Saves Water & Fertilizers Compare to Conventional method by using High Density Plantation (HDP).

High Density Plantation- (HDP)

HDP is one of the improved production technologies to achieve the objective of – enhanced productivity of fruit crops. HDP gives higher yield as well as returns/unit area due to increasing the no. of trees/unit area. It is possible by regular pruning and use of bio regulators for develops proper plant architecture and annual canopy management.

High Density Plantation (HDP) is given to the 3 farmers of the three districts

Padhiyar Harkhaben Dahyabhai from Jambusar District (Latitude – 22.178992 & Longitude – 72.866940) have used this method for her horticulture orchard and she found following benefits of this method.

- Increases yield per unit area and improves fruit quality.
- Reduces labour cost resulting in low cost of production.
- Enables the mechanization of fruit crop production.
- Facilitates more efficient use of fertilizers, water, solar radiation, fungicides, weedicides and pesticides'.

It includes planting trees as close as possible in the same area which not only saves space, but the planted saplings also support each other in growth and block sunlight from reaching the ground, thereby preventing the growth of weed.

Climate Related Services

India has Sustainable Agriculture growth.

- Agriculture sector Agriculture, Animal Husbandry and Fisheries are now facing unprecedented challenges in the new arena of Climate Change.
- Adverse weather events are increasing with intensity. This includes delayed or lack of rain, cyclones, unseasonal rains, dust storms, locust attacks, heat waves, increased temperature.
- AGROMET weather advisory has been a great help. It is noticed that farmers who have received this along with agro advisory have sustained their crops and saved their animals and fishermen have saved their lives.
- In reality local advisory is missing. Only a few KVKs transfer – issue guidance to blocks and villages.
- Country has advanced digital systems, drones and AI in Agriculture.
- It is now feasible to reach out to Farmers at Village level.
- This needs an integrated approach of merging Satellite Imagery Drones and AI for weather advisory at Village level.
- ICAR, State Agricultural Universities and KVKs have to play a very crucial role in reaching out to farmers by using Digital Technology.
- It must be realized that local level information and guidance of climate impacts and agricultural practices to be adapted is the key to sustainable growth in Agriculture.
- Adverse weather events are increasing with intensity. Temperature is also rising.
- This is creating farm distress – as cyclones, heavy rain episodes, unseasonal rains or locus attack they lose crops. In the event of cyclone or drought their source of livelihood lands get washed/dried out – cattle lost.

- Weather forecasting followed by Agro – advisory has prevented such losses – and farmers who have received it have sustained their income.
- Hence key to sustainability in the arena of climate change is regular weather advisory along with Agro, Animal Husbandry and Fisheries advisory throughout the year.
- This needs to be given every month – it is not only for adverse events but regular working in farms, managing cattle or going out for identified fish catch areas.
- The ICAR needs to instruct all KVKs to provide this at block level – block-wise to the ATMA team to make it available to each village panchayat.

Satellite Imagery – Drones - AI

Satellite imagery has become a game-changer for the agriculture sector, enabling a data-driven approach to farming known as precision agriculture. Here's how it's used:

- Measuring Carbon Sequestration by Agriculture – Vegetation – Seaweed. Identifying species which absorb maximum CO_2.
- *Monitoring Crop Health:* Satellites provide high-resolution images of fields, allowing farmers to identify issues like nutrient deficiencies, pest infestations, and diseases early on. This early detection enables timely interventions to minimize crop damage and ensure optimal yield.
- *Predicting Yield and Growth Stages:* By analyzing changes in spectral reflectance over time, satellite imagery can predict crop growth stages and potential yield. This information helps farmers plan for harvest, manage resources effectively, and make informed decisions about irrigation and fertilization.
- *Optimizing Resource Use:* Satellite imagery helps pinpoint areas within a field with specific needs. This enables farmers to apply fertilizers, pesticides, and water more precisely, reducing waste and optimizing resource use.
- *Crop Mapping and Land Use Analysis:* Satellite imagery can be used to create detailed crop maps, identifying the types and areas of crops planted across a region. This information is valuable for agricultural planning, resource allocation, and market analysis.

Satellite imagery plays a vital role in modern fisheries management. Here are some of its important applications:

- *Habitat Mapping and Monitoring:* Satellite imagery can be used to map and monitor essential fish habitats, such as coral reefs, seagrass meadows, and mangroves. This information is essential for fisheries management as it helps to identify areas that are important for fish spawning, feeding, and nursery grounds.

- *Fish Stock Assessment:* Satellite imagery can be used to estimate the size and distribution of fish stocks. This information is used to set catch limits and ensure the sustainability of fisheries.
- *Detection of Illegal Fishing:* Satellite imagery can be used to detect illegal fishing activity, such as fishing in closed areas or using banned gear. This information can be used to enforce fisheries regulations and protect fish stocks.
- *Monitoring of Environmental Changes:* Satellite imagery can be used to monitor environmental changes that can affect fisheries, such as sea surface temperature, chlorophyll concentration, and ocean currents. This information can be used to develop fisheries management strategies that are more responsive to environmental changes.

Drones are revolutionizing the agriculture industry by providing a new level of efficiency, precision, and data-driven decision making for farmers. Here are some of the key applications of drones in agriculture:

- *Crop Monitoring:* Drones equipped with high-resolution cameras and multispectral sensors can capture detailed images of crops, allowing farmers to identify potential problems such as nutrient deficiencies, pest infestations, and irrigation issues early on. This enables them to take timely action to improve crop health and yields.
- *Field Mapping and Analysis:* Drones can be used to create accurate maps of fields, including data on soil conditions, elevation, and drainage patterns. This information can be used to optimize fertilizer and water application, as well as improve overall farm management practices.
- *Precision Spraying:* Drones can be equipped with spraying systems to apply pesticides, herbicides, and fertilizers with pinpoint accuracy. This reduces waste, minimizes environmental impact, and protects the health of farmers by limiting their exposure to chemicals.
- *Seed Planting:* Drone-based planting systems are being developed to automate the seeding process, especially for cover crops and in challenging terrain. This can improve planting efficiency and uniformity, leading to better crop establishment and yields.
- *Livestock Management:* Drones can be used to monitor livestock herds, track their movement, and assess their health. This information can be used to improve grazing management, prevent disease outbreaks, and ensure the well-being of animals.

The use of drones in agriculture is still in its early stages, but it has the potential to significantly transform the industry. By providing farmers with new tools for precision agriculture, drones can help to improve crop yields, reduce costs, and minimize environmental impact.

Climate change poses a major threat to our planet, and a combination of satellite imagery, drones, and artificial intelligence (AI) is proving to be a powerful tool in combating it. Here's how these technologies work together:

- Satellite imagery provides a long-term view of Earth, allowing scientists to monitor changes in climate patterns, such as rising sea levels, melting glaciers, and deforestation.
- Drones can capture high-resolution images and collect data at specific locations, enabling researchers to study the impacts of climate change on local ecosystems and communities.
- AI can analyze vast amounts of data from satellites and drones, identifying trends and patterns that would be difficult for humans to detect. This allows for more informed decision-making regarding climate change mitigation and adaptation strategies.

Country is already using all three modern technologies, namely Satellite Imagery, Drone and AI. In fact, the use of Satellite Imagery has been in use since the late seventies. Drone and AI introduced recently. It is suggested to develop a model project for integrating all three.

Exports and Imports

We are emerging as a major exporter of Agriculture products including Food-grains. With second largest arable land, adaptive farmers and largest pool agriculture Scientists, We can supply world almost every agriculture produce and products. But time and again due to domestic inflation, a ban on several items like wheat, rice onion etc has brought down exports. E.g. Our exports totaled 53.15 billion in 2022–23 which came down to 48.82 billion in 2023–24 (US dollar). While in imports there was marginal decline of 7% in 2023–24 in edible oils, that of pulses almost doubled.

The policy matter relates to ban on exports and easing of restrictions on imports by doing away with import duties on edible oils. Both need review. Important matter to realize is that with some decline in production urban inflation pickup and with increase in production, prices to producers - farmers decline. This gives opportunity to push Food in politics and results are agitation. It is important to realize that both less or more production is predicted or are predictable and Agriculture and Food administration with timely response and action can mitigate the recurring similar situation.

Urban Agriculture

- This is a neglected area but a key place for reducing GHG emissions.
- Currently, trees – plants – flowers. Trees planted without keeping a view of its capacity grow in urban climatic environments.
- Further there is a need to focus on selection of trees which absorb maximum CO_2. ICAR – Agro-Forest Research Institute has identified such species.
- ICAR should make this available to all state Governments with detailed agro practices to be followed.
- At urban centers – there are no agro service centers or information outlets. This results in urbanities bringing expensive plants with high mortality and waste of resources and opportunity to absorb CO_2.

Farmers

- This is in order to achieve the goal of 'Atmanirbhar' – it is the farmer who is key to its success. Farmers and their association need to be made aware about action they have to take to increase their income.
- Alert – he must keep himself up-dated climate change, and the danger of health hazards.
- About benefits available under existing and new schemes of Govt.
- Smart with the ability to Adapt/adopt new technology: Climate Smart Farming, Good Agricultural Practices (Best Management Practices), Drip and Sprinkler Irrigation and Solar appliances.
- Knowledgeable – Keep track of market price of inputs and agri-produce including that of MSP and select his crop based on its Soil Health analysis – selection high value and low volume crop. Take benefit from all Govt. Schemes. Sell crops – directly to APMC, MSP or trader/industry whoever offers the maximum price.
- Able to secure/resort to multiple sources of income – livestock, solar – sale of excess energy generated, bee keeping, handicraft etc. and take employment under MGNREGA.
- Sensitive to Safety – safety advice given by Health Deptt. for himself, family, workers. For safety of crops follow weather and agro-advisory and insure the crop.
- Able to Add – value and get better price by cleaning, grading, sorting, properly packing and selling where the highest price is available.

- For all above Capacity Building is needed with modern media along with traditional radio and tv networks.

Opportunity in Climate Change

- Climate change is causing adverse impacts. Adverse events like cyclones with floods and heavy rains, single heavy rain episodes, droughts, hot and cold waves, dust storms, etc. with increasing temperature. All this is affecting normal life and livelihood – more particularly in rural areas.
- This is affecting the agriculture sector, families, organisations and public governance – government at all levels from local bodies to state and central. This requires change in existing- established practices. This is a disruption that brings about change under compulsion and something that was needed but not brought about by "*Ho ta hai chalta hai*" attitude at all levels. But it provides new openings also.
- First foremost it establishes the importance of agriculture for food security and more importantly as a natures technology for absorbing CO_2 by its photosynthesis process. It is emerging as a major Mitigation too.
- Secondly within agriculture neglected but drought and stress resistant crops like Millets are getting noticed and now encouraged animal breed – indigenous breeds which are capable of surviving in higher temperature and sea vegetation like seaweed and like.
- Third is efficient use of water and energy. farmers will have to use moisture meters and drip irrigation and adopt roof top solar panels for pumping. Even housing will have to be redesigned and use of bamboo will become popular. Urbanites like to shift to stay in smart villages to have a cooler environment that will open up huge local level employment opportunities.
- Further living style will change. Cloths, footwear will undergo change. This opens opportunity for light cotton handmade clothes. Traditionally we had this.
- Khadi and village industry will get boost. Further on industrial side, high velocity winds requires strengthening of Electrical transmission polls, a huge program of salinity ingress prevention walls, so on and so forth.
- The public transport both on road and rail can be electrical with back up of solar energy. In both solar and wind energy system will produce more. In fact all community ponds, canals can be covered by solar panels which will also reduce huge evaporation losses.

- This is some ideas which have scope to identify many more opportunities.
- NCCSD has monitored the outcome with the help of Giotag report and make a impact Assessment. Detail case studies are as under.

Conclusion

We have not dealt with program implementation, and which is key to success. There could be a Nodal Department – Agriculture Ministry and monitoring in Niti Aayog, that will make this happen.

Our experience of Krishi Mahotsav – introduced by Hon'ble Prime Minister as the Chief Minister of Gujarat in the year 2004 is very briefly shared below. This transformed Gujarat agriculture. It had a minus growth rate, which jumped to 11 per cent per annum.

- That involved 17 Departments connected with agriculture.
- A single Government Resolution laying out all activities to be performed by each department – including that by District – Taluka and Village Level Functionaries.
- Day to day monitoring and feedback.
- The Chief Minister himself reviewed it on a day-to-day basis.
- Agriculture Department was declared – Nodal Department.
- A massive water harvesting program to cover all villages with ponds and check dams.
- Free input kit for 15 poorest farmers of each village.
- Guidance to farmers to select crops which can be sustained by his soil, based on soil health analysis.
- All Officers and Ministers required to visit villages – including Scientists and DM & DDO for village level interaction.
- Out-come was staggering - increase growth rate from Minus to 11% per annum.
- NITI Aayog may lay out overall Action Plan and demarcate action at central level by Departments and ICAR – CSIR & State Governments.
- It is important is restructuring of ICAR – making its institutes accountable and re-organization in the Ministry of Commerce and Ministry of Water Resources.
- Development is a process. Countries move from one phase to another.
- India started with acute poverty, famines and scarcity.
- Moved to pre-developing stage and developing stage, from Green revolution to White revolution and current Blue Revolution.

- It is now under transition to becoming a Self-Reliant – Atmanirbhar, a developed country.
- Prime minister has called it AMRUT KAAL, with all-around prosperity and poverty of none.
- Country has the second largest cultivable land, manpower and immense sun, water, and sea resources. It has adaptive farmers and a large pool of agriculture scientists and a competent public leadership.
- It is fortunate to have a very strong, visionary, and dynamic leadership of Prime Minister Shri Narendra Modi.
- Country is capable of making this happen.

The Critical Path

- Involve and make responsible Panchayati Raj Institution and leadership.
- Re-orient and re-designed water distribution and conservation policy and monitor losses and fix accountability.
- Rationalize solar purchase policy and expand it on canal and community ponds.
- Organize Market – remove control of Traders on Agri produce-Replicate Milk Marketing.
- Create integrated R&D centers.
- Introduce Micro level Block level planning with nexus of food, Water-energy – by using digitalization.
- Make Agricultural University to reach out farmers with Agro Advisory followed by weather Advisory at village level.
- Solving Sharecroppers Problem.
- Prioritize sea weeds, Bamboos energy sale by farmers, use Cultivable wasteland and wetland.
- Identify Rural poor – support them at individual level.
- There is no need to Introduce new Schemes. We have a well laid out program.
- Ensure that food wastage is reduced by food collection and distribution system – Collection at village level.

Water Security in Agriculture—Identifying Advanced Technology Confluence

V.K. Saraswat[1]

Introduction

Agriculture, the backbone of global food production, stands at the cusp of water management challenges and opportunities. As the United Nations predicts a 50% increase in agricultural production by 2050, concerns about water scarcity, inefficient irrigation practices, and environmental impacts come to the forefront. To support a daily diet of 2,800 kilocalories, 2,000 to 5,000 liters of water per person per day, the United Nations anticipates a 30% increase in global water withdrawals for agriculture by 2050. Presently, 2.4 billion people face water stress, and nearly 40% of global crop lands experience water scarcity. Small-scale farmers, responsible for a significant portion of global food production, grapple with water challenges exacerbated by climate change.

Indian Water Landscape

According to the Indian Agricultural Statistics Report 2022, the agriculture sector employed over 50% of the Indian workforce, particularly those in rural areas and contributed 18.8% to the country's GDP. However, India, with 18% of the world's population and only 4% of its water resources, confronts a skewed dynamic. Agriculture, consuming a staggering 90% of the country's water, far surpasses the global average of 72%. The nation's per capita water availability has declined from 5000 cubic metres (m^3) per annum in 1950 to around 2000 m^3 and is projected to decline to 1500 m^3 by 2025, leading to far less water availability for agriculture.

Present Scenario of Concern

Factors such as over-irrigation, inadequate water distribution, climate change, population growth, soil erosion, and pollution exert immense

[1] Member, NITI Aayogj, Government of India, New Delhi.

pressure on agricultural water resources. Conventional crops, lacking micro-irrigation adoption, neglected storage capacities and poor water infrastructure compound the challenges and contribute to water scarcity. Coupled with these issues, farmers in many regions face increasing competition from non-agricultural users due to rising urban population density and water demands from the energy and industry sectors.

Resolving future water challenges in agriculture requires a systematic reconsideration of water management and reassessment of overall water resources management and water policy in agriculture. Existing institutional policies, both at the national and international levels, cannot adequately address agricultural water management to ensure future water and food security.

In the view of agricultural framework, both supply- and demand-side measures are crucial for realizing water management, based on the effectiveness and ease of implementation of strategies. Demand-side measures like structural and operational changes (e.g. replacing inefficient water pumps, using drip irrigation, laser land levelling), economic measures (e.g. financial incentives on reducing water waste in irrigation), and provision of training and educating farmers towards water-saving technologies – may prove to be instrumental.

On the other hand, supply-side management techniques can be expensive and, thus, their implementation may face financial barriers in developing countries. Yet, measures like implementing small farm-sized dams, growing crops based on water requirements and water availability, are commendable proceedings.

Key Interventions

A multitude of interventions offers a comprehensive framework for sustainable water management.

- ***Technical Interventions:*** Bringing crop & livestock closer together with manure management technologies
- ***Social Interventions:*** Combining scientific & practical knowledge through education
- ***Economic Interventions:*** Accounting for cost-effectiveness in solutions
- ***Policy Interventions:*** Consider synergies and trade-off between agricultural practices and the environment in spatial planning
- ***Environmental Interventions:*** Building on nutrient recycling & ecological resilience

What are the Agricultural Best practices for Water Conservation

Drip irrigation, irrigation scheduling, dry farming, compost and mulch, cover crops, capturing and storing water, drought-resistant crops, rotational grazing, conservation tillage, and organic farming are vital practices for water conservation. This also includes practices such as rainwater harvesting, desalination, Greywater Recycling, and Aquifer Recharge. Equally important in water quality protection are nature-based solutions (NBS), which use nature-driven processes to address socio-environmental challenges. By mimicking nature, these solutions create a symbiotic relationship between human activities and the environment, enhancing ecosystem health and resilience.

Advanced Technology Approaches

To realize sustainable water management strategies, leveraging the potential of advanced technologies will be the best route. A suite of technologies can enhance water efficiency – ensuring that each unit of water used in agriculture generates the maximum possible value.

AI-Driven Precision Agriculture

Integrating AI-driven technologies like machine learning, computer vision, and remote sensing allows farmers to collect and analyses large volumes of data in real-time. These data-driven insights help optimize farming practices, including irrigation, fertilization, and pest management. As a result, AI-driven precision agriculture can significantly improve water efficiency and contribute to sustainable farming practices.

Smart Irrigation Systems

These systems use various sensors, such as soil moisture sensors and weather data, to monitor the water requirements of crops. Machine learning algorithms then process this data and provide real-time irrigation recommendations. This approach enables farmers to apply water only when and where needed, reducing water waste and improving overall efficiency.

Predictive Analytics and Crop Modeling

AI models can predict crop water needs and growth stages by integrating historical data, weather patterns, and crop-specific information. These insights are used to optimise irrigation schedules and adapt to changing environmental conditions.

Satellite and Drone-Based Remote Sensing

Remote sensing technologies, such as satellite imagery and drones (UAVs—unmanned aerial vehicles), are increasingly used to monitor crop health, soil moisture, and other environmental parameters. AI algorithms can analyse these large datasets to identify patterns and trends related to water use efficiency. For example, remote sensing can detect areas of over-irrigation or water stress, allowing farmers to make targeted adjustments to their irrigation practices.

Internet of Things (IoT) in Agriculture

IoT applications in agriculture involve the integration of smart devices and sensors connected through the internet. In the context of water management, IoT facilitates real-time monitoring of water infrastructure, automated irrigation systems, and data-driven decision-making. This interconnected ecosystem enhances efficiency and reduces water wastage.

Other Emerging Technologies

Nanotechnology for water purification, machine learning algorithms for predictive analytics, and the integration of block-chain for transparent water transactions are some of the emerging trends that hold promise for the future.

Moreover, continued research and development in the field are giving rise to innovative technologies.

Circular Economy Approach

A circular economy in agriculture implies a regenerative system that reduces resource input, waste, and environmental degradation. The primary goal is to close the loop on resource use, creating a system where waste is minimized and resources are recycled and reused, the same can be replicated with water practices.

Start-up Contributions

Start-ups, worldwide, are playing a pivotal role in driving innovation in agricultural water management. Companies are actively developing cutting-edge solutions, from AI-powered irrigation controllers to satellite-based precision farming tools. Their contributions are fostering a competitive market and accelerating the adoption of advanced technologies. India is prominently positioned in the global startup heat-map for water management (Figure 1). Strategies involving technology and innovation investments, rainwater harvesting, spate irrigation, and renewable energy pumps showcase India's commitment to addressing water challenges.

Market Projections

The market for advanced technologies in agricultural water management is witnessing a robust growth. According to the latest projections, the precision farming market is estimated to reach around USD 34.01 billion by 2032 with a CAGR of 13.30%, whereas the smart agriculture market is expected to grow from USD 16.2 billion in 2022 to USD 30.4 billion in 2030 with a CAGR of 9.4%, in the Asia-Pacific region alone. In India, smart agriculture market growth is anticipated to progress at a CAGR of 13.38% during 2022–2028, and is expected to garner a revenue of USD 886.21 million by 2028.

Figure 1: Start-up Global Presence (*Source:* Stratus Insights)

The national market is poised for rapid expansion, driven by government initiatives, increasing awareness, and the need for sustainable farming practices.

Government Initiatives

Government of India have initiated targeted schemes to address the pressing needs of the agricultural sector - *ATAL Bhujal Yojana*, National Mission for Sustainable Agriculture (NMSA), *Pradhan Mantri Krishi Sinchai Yojana* (PMKSY), Accelerated Irrigation Benefit Programme

(AIBP), Command Area Development & Water Management (CADWM).

These schemes are intended towards improved utilization of water resources, water distribution mechanism, micro-irrigation practices, region-based targeted irrigation technology and end-to-end solution to agricultural practices.

At this juncture, policy frameworks that incentivize the use of precision agriculture, promote the development of smart infrastructure, and provide financial support for technology adoption are essential for water management practices. Government collaboration with the private sector can further accelerate this transition. Moreover, initiatives to ensure that farmers and stakeholders have the necessary skills to leverage advanced technologies is crucial. Government agencies, academic institutions, and private enterprises should collaborate to explore new technologies, improve existing ones, and address challenges unique to the Indian agricultural landscape.

Conclusion

Achieving water security in the agricultural sector is a multifaceted challenge that demands a holistic approach. Advanced technologies present a transformative opportunity to address water scarcity issues and enhance agricultural productivity, by embracing the prospects of AI and ML, IoT, digitalization and circular economy principles, ensuring a more sustainable and water-secure agricultural landscape. As we navigate towards the Sustainable Development Goals (SDGs) for 2030, integration of these technologies into mainstream agricultural practices becomes not just an option but also a necessity for a water-secure and food-sufficient India.

Disclaimer: *Views, thoughts, and opinions expressed in the article belong solely to the author.*

Empowering Women in Agriculture: Shifting Dynamics and Gender Balance

Neelam Patel[1]

Introduction

India commemorates its 78 years post-independence with the vibrant fervour of '*Azaadi Ka Amrit Mahotsav*' while championing women's empowerment through the rallying cry of 'Empowered Women, Empowered Nation'. In an economy deeply rooted in agriculture, where 80% of the female workforce is engaged (GoI, Census, 2011), it is imperative to accord farm women equal significance and contemplate their empowerment. Particularly in rural areas, women demonstrate a commendable workforce participation rate of 41.8 %, outstripping their urban counterparts at 35.31% (MOSPI, 2017) as depicted in Figure 1. Women exhibit substantial involvement in primary crop production (75%), horticulture (79%), post-harvest activities (51%), as well as animal husbandry and fisheries (95%) (Jadav, 2022). Table 1 illustrates that active farm women spend 5 to 9 hours daily on agricultural activities. Despite constituting 33% of the agricultural labour force and 48% of self-employed farmers in India, they have long been marginalized, overlooked, and relegated to the status of 'invisible farmers'.

Table 1: Involvement of Women in Different Crop Production Activities

Activity	*Involvement (%)*
Land preparation	32
Seed cleaning and sowing	80
Inter cultivation activities	86
Harvesting-reaping, winnowing, drying, cleaning, and storage	84

Lately, shifting dynamics of the agricultural sector have highlighted female involvement across diverse occupations such as farming, entrepreneurship, and rural labor. This shift is partially driven by male

[1] NITI Ayog, Government of India, New Delhi, India.

migration to urban centers, as highlighted by the Indian Economic Survey of 2017–18, signaling the onset of an era characterized by the feminization of agriculture. The female-to-male ratio in agriculture in rural areas has increased from 1.1 to 1.36, as depicted in Figure 2, representing a notable shift towards a more balanced gender representation in the sector. Furthermore, mounting evidence has acknowledged that forces like climate change disproportionately burden women (Goh, 2012) (Habtezion, 2013). The intersection of poverty and vulnerability to climate impacts can push them further into a vicious cycle of poverty and deprivation. Therefore, it's of paramount importance to prioritize women's empowerment and integrate gender perspectives into agricultural strategies. This, ultimately, calls for a concerted effort towards gender-sensitive innovations to foster farm women's financial and holistic empowerment (ADBI, 2023).

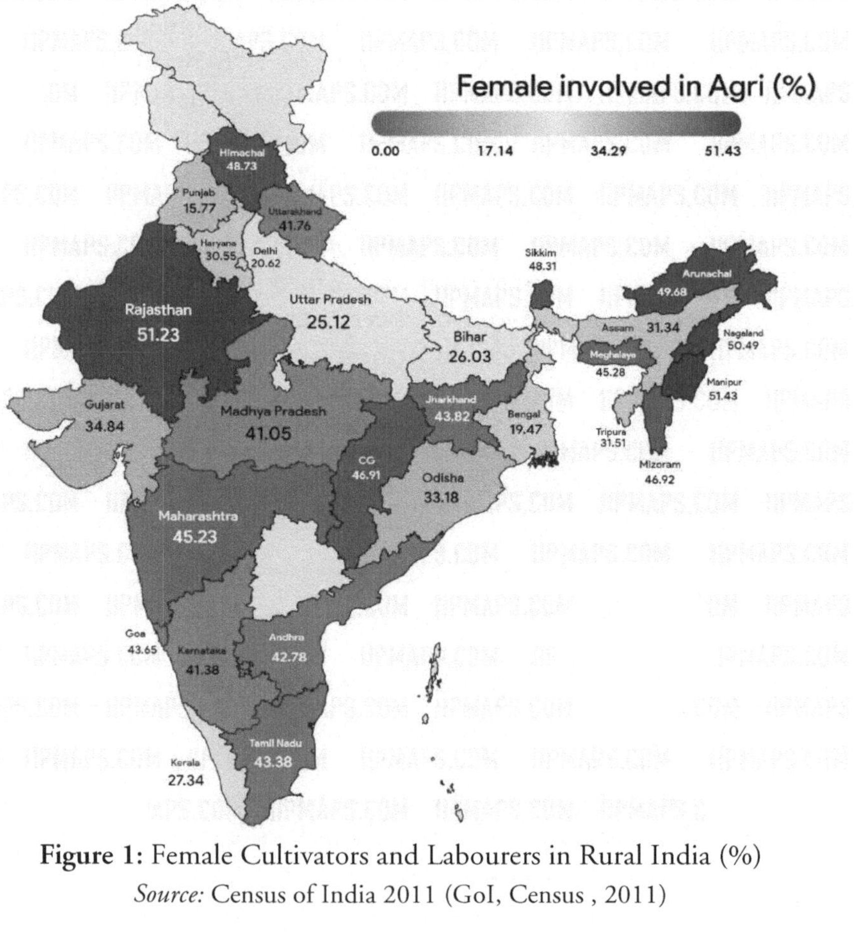

Figure 1: Female Cultivators and Labourers in Rural India (%)

Source: Census of India 2011 (GoI, Census , 2011)

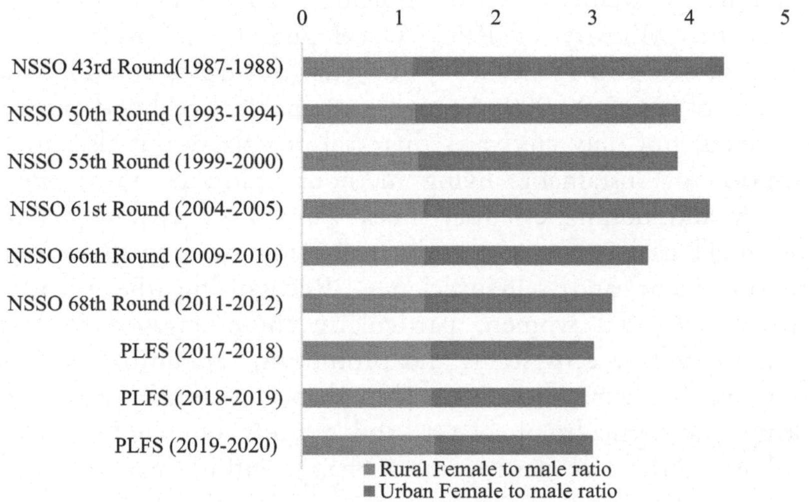

Figure 2: Distribution of Workers in Agriculture in India (1987–1988 to 2019–2020)

Reports indicate that factors such as education, agricultural knowledge enhancement, access to information and marketing capabilities, technological advancements, and adequate support and resources significantly influence women's empowerment (Nath & Athinuwat, 2020), (ADBI, 2023) (Woods, 2022).

Governmental Efforts for Empowering Women in Agriculture

Considering this, the Government of India has undertaken several steps to empower farm women. National Policy on Farmers, 2007 has included "mainstreaming the human and gender dimensions in all farm policies and programs" as one of the major policy goals. Mainstreaming of gender concerns is being addressed by (i) earmarking 30% of funds for women under various major schemes/programs and development interventions; (ii) taking pro-women initiatives to help women derive the benefits of beneficiary-oriented components of various programs/schemes/missions. Focus is also being given to the formation of women Self Help Groups (SHGs), capacity-building interventions, linking them to micro-credit, enhancing their access to information, and ensuring their representation in decision-making bodies at various levels. In alignment with governmental endeavors, initiatives like the Mahila Kisan Sashaktikaran Pariyojana (MKSP), led by the Ministry of Rural Development, systematically invest in bolstering women's engagement and efficiency in agriculture, ensuring its enduring sustainability. Additionally, Lakhpati Didi, another initiative under the Ministry of Rural Development, empowers farm women affiliated with Self-Help Groups (SHGs) to surpass an annual household income of Rs. 1,00,000 through sustainable livelihood practices. This endeavour not only cultivates financial prosperity but also promotes the adoption of sustainable living methods. Through imparting financial literacy and honing essential skills, women are encouraged to explore entrepreneurial ventures, thereby reshaping rural socio-economics with empowerment and self-sufficiency. Recognizing the pivotal role of empowered rural women, particularly those engaged in farming, in fortifying village ecosystems and promoting agricultural sustainability is paramount. Their efforts not only enhance personal well-being but also fortify the agricultural sector and ensure household food security, establishing them as silent agents of change within their communities.

Empowering Through Access to Assets

Through the Mission on Seed and Planting Material (SMSP), women farmers benefit from subsidized rates on planting materials. Similarly, the Sub-Mission on Agricultural Mechanization (SMAM) strives to boost productivity and efficiency in agriculture by promoting the adoption of agricultural machinery. In a concerted effort to empower women farmers, SMAM offers a higher subsidy rate and allocates funds specifically for acquiring agricultural machinery. This initiative aims to enhance the farm power status of women farmers, enabling them to adopt modern farming techniques and improve overall productivity.

Empowering Through Group Formation and Access to Technology

Expanding into dairy development, the Ministry of Rural Development empowers farm women through the Deendayal Antyodaya Yojana-National Rural Livelihoods Mission (DAY-NRLM), supporting over 81,000 women farmers in dairy value chain development. Likewise, the Dairy Entrepreneurship Development Scheme, administered by the Department of Animal Husbandry, Dairying & Fisheries, bolsters small dairy farms and prioritizes women in self-help groups, cooperatives, and producer companies.

Innovative schemes like the Namo Drone Didi under the Sashakt Nari-Viksit Bharat program promote digital agriculture among women by providing agricultural drones and training, benefiting 15,000 women-led

Self-Help Groups (SHGs) for tasks such as spraying, sowing, and crop monitoring. Additionally, programs like Agri-Clinics and Agri-Business (AC&ABC) aim to transform unemployed youth into Agripreneurs, with women receiving higher subsidies. Furthermore, platforms like Startup India offer various schemes and programs aimed at fostering women's entrepreneurship, including state-level policies and central government initiatives showcased on the portal. The Biotech-Krishi Innovation Science Application Network (Biotech-KISAN) Program, launched by the Department of Biotechnology (DBT), aims to offer scientific solutions to farmers in the northeastern region. Its objective is to connect innovative agricultural technologies with small and marginal farmers, particularly focusing on women farmers in the area (PIB, 2021).

Empowering through Financial Inclusion

India has embarked on a significant endeavour to expand banking services to rural regions. Through the Pradhan Mantri Jan-Dhan Yojana (PMJDY), financial inclusion and accessibility to banking have been greatly enhanced, particularly empowering rural women to engage in economic activities. The Jan Dhan campaign ensures that rural women have affordable access to financial services such as banking/savings and deposit accounts, remittance, credit, insurance, and pension. These measures promote transparent transactions and timely access to financial services, including direct benefits transfer (DBT) under various Government of India schemes. In the seven years since its inception, over 43.04 Crore accounts have been opened nationwide, with 55.47 percent (23.87 crore) held by women and 66.69 percent (28.70 crores) located in rural and semi-urban areas (GoI, Pradhan Mantri Jan-Dhan Yojana (PMJDY) – National Mission for Financial Inclusion, completes seven year of sucessful implementation, 2021).

Financial inclusion has played a crucial role in enabling the rural population to withstand the challenges posed by the COVID-19 pandemic by ensuring uninterrupted access to financial assistance. Complementary initiatives such as the Pradhan Mantri MUDRA Yojana (PMMY), Stand-Up India Scheme, and Prime Minister's Employment Generation Programme (PMEGP) further bolster financial empowerment and entrepreneurship development among rural women. A collective total of over 9 crore women have benefited from Mudra and Stand-Up India schemes (Ashish Kumar, 2019).

Conclusion

Rural women are the backbone of India's agricultural sector, yet they often face hurdles in reaching their full potential. We must empower these vital stakeholders to cultivate a more prosperous and equitable New India. By ensuring equal access to resources, technology, education, and healthcare, we can unlock the immense potential of rural women. Land ownership rights and skill development programs will further equip them to thrive. This holistic approach will not only boost agricultural productivity but also foster a generation of empowered and self-sufficient women, contributing significantly to a stronger nation. The government's initiatives in education, financial aid, and fostering collaboration among women are commendable steps in the right direction. These programs empower farm women, strengthening the agricultural sector and paving the way for a more secure and prosperous future for all.

References

[1] ADBI (2023). Women's Economic Empowerment as a Pathway Toward Sustainable and Inclusive Development in India. Tokyo, Japan: Asian Development Bank Institute. Retrieved from https://www.adb.org/sites/default/files/publication/917246/adbi-women-s-economic-empowerment-pathway-toward-sustainable-and-inclusive-development-india.pdf

[2] Ashish Kumar (2019). Women's empowerment under Modi Government. Retrieved May 16, 2024, from https://myvoice.opindia.com/2019/03/womens-empowerment-under-modi-government

[3] Deshpande, A. and Kabeer, N. (2021). "Norms That Matter: Exploring the Distribution of Women's Work betweenIncome Generation, Expenditure-Saving, and Unpaid Domestic Responsibilities in India. WIDER Working Paper Series. Retrieved from https://www.econstor.eu/handle/10419/248344%0A https://ideas.repec.org/p/unu/wpaper/wp-2021-130.html

[4] Deshpande, A. and Singh, J. (2021). Dropping Out, Being Pushed Out or Can't Get in? Decoding Declining Labour. IZA Discussion Paper No. 14639. Retrieved from https://dx.doi.org/10.2139/ssrn.3905074

[5] Goh, A.H. (2012). A literature review of the gender-differentiated impacts of climate change on women's andmen's assets and well-being in developing countries. International Food Policy Research Institute, CAPRi Work. Retrieved from https://doi.org/10.2499/CAPRiWP106

[6] GoI (2011). Census. Census of India, Government of India.

[7] GoI (2021). Pradhan Mantri Jan-Dhan Yojana (PMJDY) - National Mission for Financial Inclusion, completes seven year of sucessful implementation. Ministry of Finance.

[8] Habtezion, S. (2013). Overview of linkages between gender and climate change,. Policy Brief, United Nations Development Programme. Retrieved from. http://www.undp.org/content/dam/undp/library/gender/Gender%20

[9] Jadav, R. (2022, 10 13). How women are doing the heavy-lifting in agriculture. Retrieved from https://www.thehindubusinessline.com/data-stories/data-focus/how-women-are-doing-the-heavy-lifting-in-agriculture/article66004569.ece

[10] Krishna, V., Padmaja, S. and Jose, M. (2023). Shifting Gender roles in Agriculture: Estimating the impact of a women-centric R&D program in India. Cereal Systems Initiative for South Asia (CSISA).

[11] MOSPI (2017). Participation in Economy. New Delhi: Social Statistics Division, Central Statistics Office, Ministry of Statistics and Programme Implementation, Government of India. Retrieved from https://www.mospi.gov.in/sites/default/files/publication_reports/Women%20and%20men%20in%20india-2017%20Mail.pdf

[12] Nath, T. and Athinuwat, D. (2020). Key factors of women empowerment in organic farming. Geo Journal, 86, 2501–2520.

[13] OXFAM India (2018). Move over 'Sons of the Soil': Why You Need to Know the Female Farmers That Are Revolutionizing. OXFAM India. Retrieved from https://www.oxfamindia.org/women-empowerment-india-farmers.

[14] PIB (2021). DBT Issues a Special Call under Biotech KISAN Programme for improving Agriculture Productivity in the North East Region. Retrieved from PIB: https://pib.gov.in/PressReleasePage.aspx?PRID=1728507

[15] Woods, Z. (2022). How Supporting Women Farmers Can Ease Asia's Food Security Crisis. Retrieved from ADB Blog: https://blogs.adb.org/blog/how-supporting-women-farmers-can-ease-asias-food-security-crisis

Technological Innovations for Sustainable Agriculture

K.D. Nayak[1] and Arvind Pai[2]

Introduction

The original slogan "*Jai Jawan, Jai Kisan*" from the 1960s has been expanded to "*Jai Vigyan, Jai Anusandan*" in recent times as India progresses significantly in science and technology. Young Indians, whether soldiers, athletes, scientists, engineers, entrepreneurs, or creators, have brought glory to our nation and contributed to its economic progress. Although agriculture remains the backbone of our economy, the nation has developed another strong sector in software and services, with efforts underway to bolster manufacturing. Our farmers and agriculturists cannot be left behind.

Our farmers are indeed the backbone of the country, aiding in achieving self-sufficiency in food grain cultivation during the Green Revolution and in milk production during the White Revolution. India boasts the second-largest arable area globally, at 1,597,000 km^2, following the United

[1] Former Director General (Micro-electronics Devices, Computional Systems and Cyber Security), DRDO, Ministry of Defence.

[2] Founder, Creative Insights, Bengaluru, India.

States. With a gross irrigated crop area of 826,000 km^2, India holds the world's largest irrigated crop area. It ranks among the top three global producers of various crops, such as wheat, rice, pulses, cotton, peanuts, fruits, and vegetables (Figure 1). Additionally, India possesses the largest herds of buffalo and cattle worldwide, making it the leading producer of milk and a significant player in the poultry industry's growth. India also stands as the world's second-largest producer of several dry fruits, agriculture-based textile raw materials, roots and tuber crops, pulses, farmed fish, eggs, coconut, sugarcane, and various vegetables. India contributes globally about 23% of tea production (ranked 2nd) and about 5% of coffee production (ranked 6th).

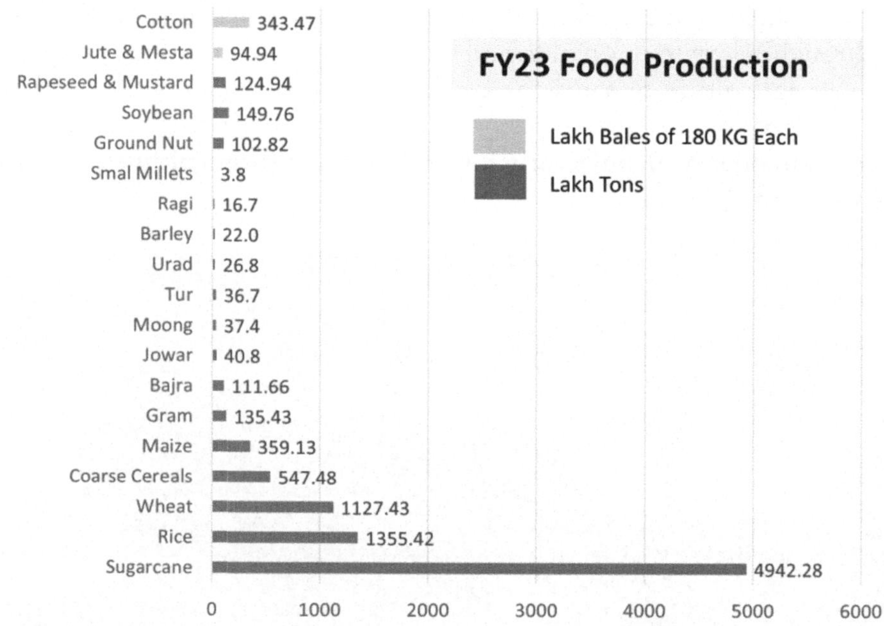

Figure 1: Food Production (FY 2023)

The Gross Value Added (GVA) of the agriculture and allied Sector during 2022–23 amounts to Rupees 2221092 Crore, with a growth rate of 3.3%, contributing approximately 21% to India's GVA. India further maintains its status as a net exporter of agricultural produce. About 54.6% of India's workforce is engaged in agricultural activities. Across India, there are 15.65 Crore agriculture holdings, with approximately 68.45% classified as marginal (<1 hectare), 17.62% as small (1–2 hectare) holdings and the remaining 13.93% exceeding 2 hectares.

Despite these green shoots, Indian agriculture and farmers face numerous challenges (Figure 2). Dependency on monsoon rains, their unpredictability due to environmental factors, scarcity of rural labour, volatility and fluctuations in agricultural produce pricing, inflation in fertilizers and seeds, lack of market access, fair prices due to connectivity deficiencies, crop losses from excessive rains or droughts leading to financial stress, and difficulty obtaining loans at fair terms are among them. The agriculture sector has not seen significant growth for decades, necessitating India to strategize to double, if not triple, the agriculture growth rate in the coming decades. Technology adoption could help address some challenges and enable higher expected growth.

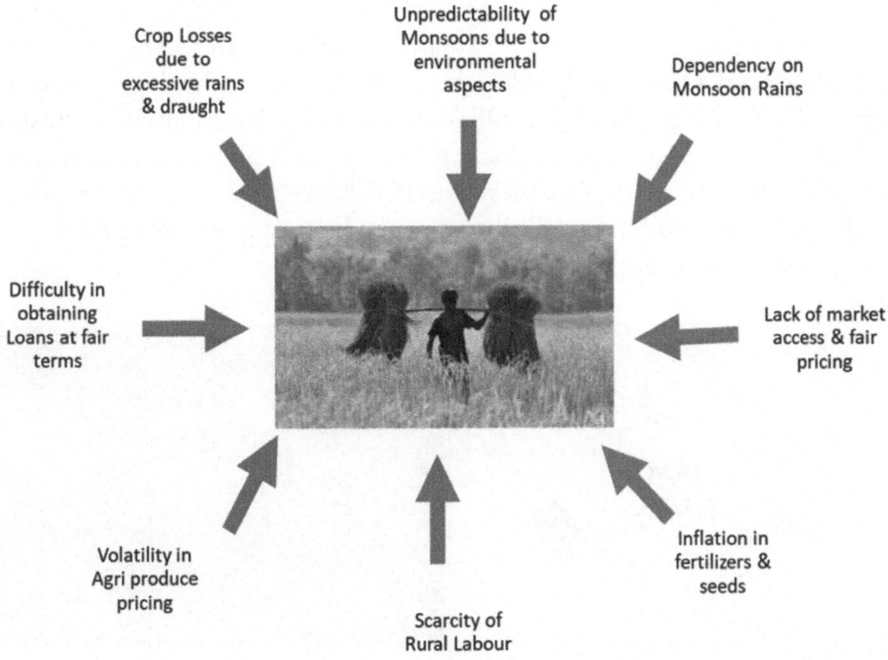

Figure 2: Challenges Faced by Indian Farmers

The Saptapadi (Saat Phere or Seven Steps) ritual symbolizes commitment, love, mutual respect, and partnership in Indian marriages. Conducted around a sacred fire, the bride and groom take seven symbolic steps together, each step accompanied by a vow, representing their journey as a married couple. To successfully merge technology with Indian agriculture, ensuring a prosperous future, we must take the following seven steps (Figure 3).

Figure 3: Seven Steps for Marrying Technology with Indian Agriculture

The integration of Jan Dhan Yojana, Aadhaar, and Mobile (JAM) technology can significantly alleviate challenges faced by Indian farmers (Figure 4). Aadhaar-linked Jan Dhan accounts enable financial inclusion, facilitate transparent and direct benefit transfers (DBT), ensuring efficient subsidy distribution for schemes like PM-Kisan and crop insurance. It also helps build financial credibility and credit rating, enabling farmers to access bank loans more easily.

Kisan Identity – Bank Accounts – Connectivity – Finance & Credit History – Farmer Loans – Crop Insurance

Figure 4: Kisan Identity—JAM Trinity

Mobile applications provide access to agricultural inputs, market prices, and weather forecasts, empowering farmers to make informed decisions. Through digital platforms, farmers can access markets directly, bypassing intermediaries and fetching better prices for their produce. Extension services delivered via mobile offer training on farming techniques and pest management.

The IIT-Bhagalpur and Bihar Agricultural University (BAU) worked together to create an innovative mobile app for farmers' benefit, with the aim of detecting agricultural diseases and related problems and finding solutions. It has been named the 'e-Nirog' mobile app (Figure 5). The e-Nirog app assists in rapidly diagnosing crop/plant ailments and locating their treatments, allowing proper pesticides/chemicals to be use to control the diseases and boost agricultural production, hence improving farmers' social and economic conditions.

Figure 5: Innovative Mobile App

Figure 6: Chitale Dairy-Tagging of Animals

Chitale Dairy in Maharashtra has successfully redefined dairy farming into a high-tech industry by automating its entire milk production and monitoring processes (Figure 6). Chitale has created a strong ecosystem of farmers and producers. Each animal is tagged with an animal tag that, when scanned, transmits unique information on each cow and buffalo, back to the Chitale Dairy data centre. Dairy operations, the feeding and breeding of animals is now monitored by computers. By automating the collection of data from each farm, they have improved animal health, leading to increased milk yield per animal. They also streamlined the efficiency of distribution channels for faster delivery of the milk products to consumers. Such a technology intervention can be scaled up across India. In future with NB-IoT enabled smart tags, animal health can be monitored remotely and be used for Animal care interventions.

Figure 7: Solar Irrigation on Farm

Technology not only enhances productivity but also fosters smart sustainable practices in agriculture. Smart irrigation systems, controlled by weather forecasts and soil moisture sensors, optimize water usage, mitigating the risk of drought and water scarcity. Conservation tillage practices, facilitated by GPS-guided machinery, reduce soil erosion and improve soil health, preserving valuable agricultural land. Moreover, integrating renewable energy sources, such as solar-powered irrigation pumps, promotes energy efficiency and reduces carbon emissions (Figure 7).

We can learn from Asia's largest Drip Irrigation Project, built at INR 3.81 billion (USD $ 52 million), the project will provide irrigation services through 2150 KMs of underground drip irrigation pipeline made up of fibre optic material to 24,000 hectares (60,000 acres) of drought-hit

Hungund in the Ramthal Marola area (Figure 8). The mega drip irrigation project will benefit more than 15,000 farmers and will also help in preventing the loss of agricultural land in India.

To use the service of drip irrigation, farmers need to deposit INR 1,300 (USD $ 17.79) per acre in the bank annually. After five years, these funds will be used for maintenance requirements. When farmers in the region opt for flood irrigation through canals, only 30,375 acres of land can be irrigated. Thanks to the drip irrigation project however, an additional 29,625 acres will be irrigated, using the same amount of water.

Figure 8: Drip Irrigation Project, Ramthal

Localized rainwater harvesting setups offer a sustainable solution to address water scarcity challenges in Indian agriculture (Figure 9). By capturing and storing rainwater on-site, these systems replenish groundwater levels, mitigate soil erosion, and provide a reliable source of irrigation during dry periods.

Figure 9: Rainwater Harvesting System

India has taken several policy initiatives to promote water harvesting. For example, Atal Mission for Rejuvenation and Urban Transformation (AMRUT), launched in year 2015, aims to promote rainwater harvesting to prevent water scarcity. Similarly, the Jal Shakti Abhiyan started in 2019, promotes the construction of rainwater harvesting systems. The National Water Policy 2021 also encourages the State Governments to create state-level implementation plans to promote the implementation of rainwater harvesting systems.

Simple techniques like rooftop rainwater harvesting, contour trenching and check dams can efficiently collect rainwater at the farm level. This localized approach reduces reliance on erratic monsoon rains and costly irrigation infrastructure, making agriculture more resilient to climate variability. Moreover, by conserving water and improving soil moisture retention, rainwater harvesting enhances crop yields, enhanced farmer incomes, and promotes sustainable agricultural practices across India.

Precision agriculture, driven by advanced technologies, is set to revolutionize farming in India. IoT sensors provide real-time data on soil moisture, nutrients, and crop health (Figure 10). Drones with advanced imaging detect pests, diseases, and water stress early, empowering farmers to optimize resources and maximize yields.

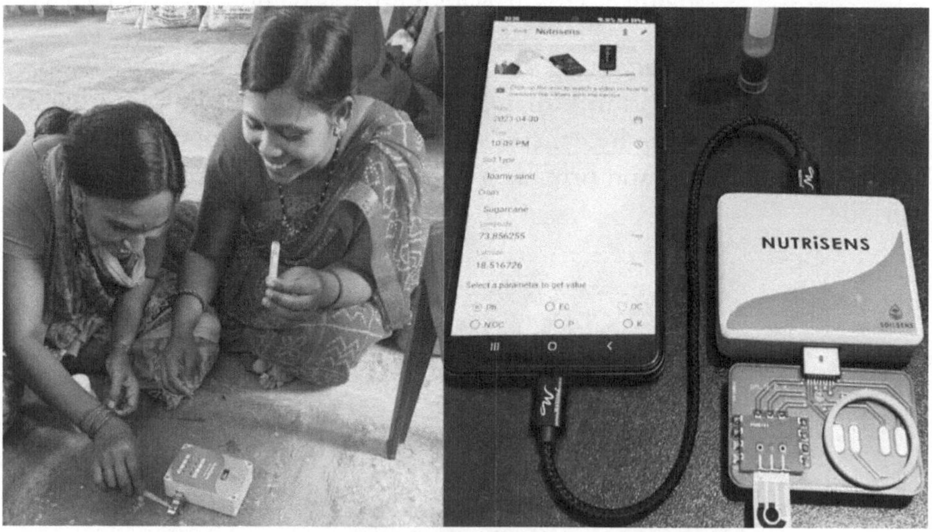

Figure 10: Sensor

Satellite-based mapping and imaging inputs offer invaluable tools for enhancing horticulture practices while also enabling farmers to earn carbon credits. By leveraging satellite technology, farmers can obtain detailed insights into their land, including soil quality, moisture levels, and vegetation health. This information allows for precise planning and management of horticultural crops, optimizing resource usage and increasing yields.

Additionally, satellite imagery can detect early signs of crop stress, pest infestations, or diseases, enabling timely intervention to mitigate losses (Figure 11). Furthermore, the data generated through satellite mapping can be used to quantify carbon sequestration efforts on agricultural land, such as the planting of trees or adoption of agroforestry practices. By demonstrating these carbon sequestration activities, farmers can qualify for carbon credits, providing them with an additional source of income while contributing to climate change mitigation efforts.

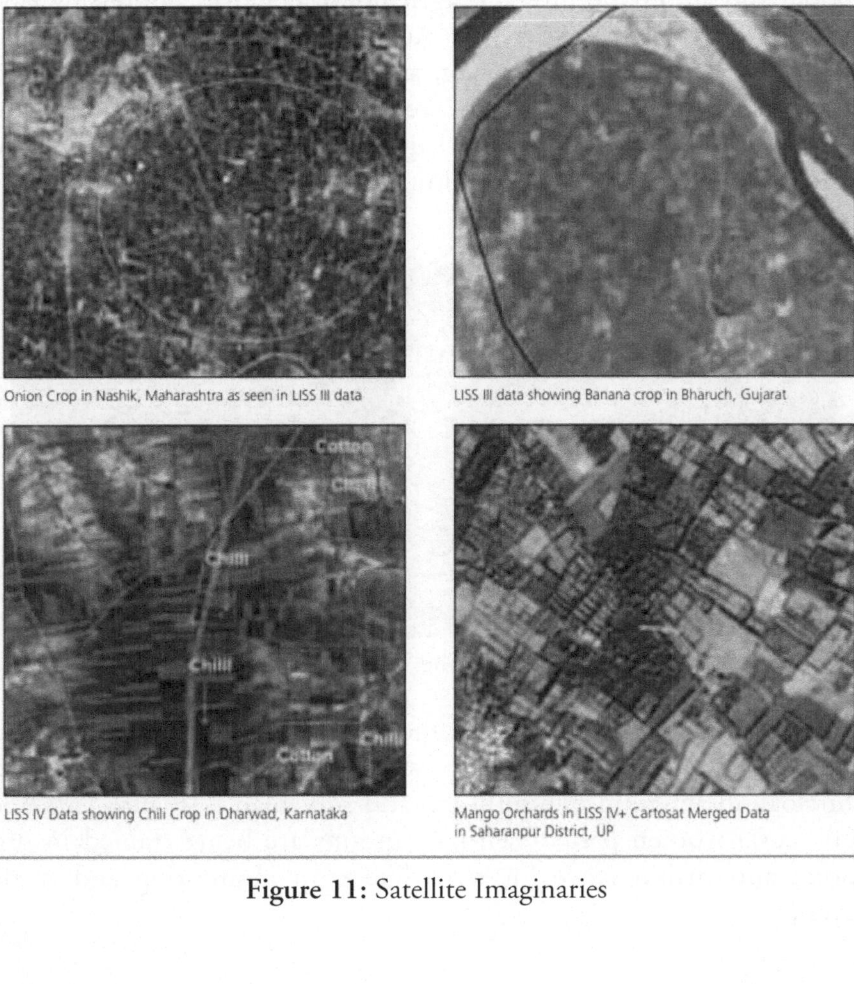

Onion Crop in Nashik, Maharashtra as seen in LISS III data

LISS III data showing Banana crop in Bharuch, Gujarat

LISS IV Data showing Chili Crop in Dharwad, Karnataka

Mango Orchards in LISS IV+ Cartosat Merged Data in Saharanpur District, UP

Figure 11: Satellite Imaginaries

Figure 12: Innovations in Agriculture

Robotics and automation reshape agricultural practices, addressing labour shortages and boosting efficiency. Autonomous vehicles with precision guidance handle planting, spraying, and harvesting (Figure 12). Robotic weeders eliminate invasive plants, reducing herbicide use and manual labour. Automated sorting and packaging systems streamline post-harvest operations, ensuring quality and minimizing waste.

Figure 13: Drone Technology Use

The "Drone Didi" program, led by the Indian government, with an initial budget of Rs, 1261 Crore empowers rural women through drone technology in agriculture, healthcare, and surveillance (Figure 13). About 15000 rural women part of self-help groups are being trained in drone piloting and maintenance for tasks like crop monitoring and medical deliveries.

Figure 14: Technology Use in Agriculture

Artificial Intelligence (AI) and data analytics have immense potential to transform agriculture into a knowledge-driven ecosystem (Figure 14). AI algorithms process data for actionable insights, from predictive weather analytics to crop disease diagnosis. Machine learning models analyse historical crop data for optimal planting schedules and crop rotations, tailored to local conditions. This personalized guidance empowers farmers, mitigating risks and boosting profitability.

Figure 15: Use of AI in Agriculture

Vertical farming offers a game-changing solution for Indian small holding farmers by maximizing space and crop yield. Farmers should identify suitable locations like rooftops, vacant land, or vertical structures (Figure 15). With government support, they can invest in necessary infrastructure like vertical growing racks, hydroponic systems, LED lights, and climate control.

According to estimates, there is about 74 Million Ton of ~22% worth food produce gets wasted due to lack of proper storage, transportation and processing facilities. Adequate storage, including warehouses and cold storage, preserves perishables like fruits, vegetables, and dairy. Investment in food processing, cold storage, Grain & Flour Mills, Oil extraction plants, and value-added processing is crucial. A robust transportation network, including roads, railways, and cold chain logistics with tracking systems, facilitates efficient movement of agricultural goods from farm to market (Figure 16).

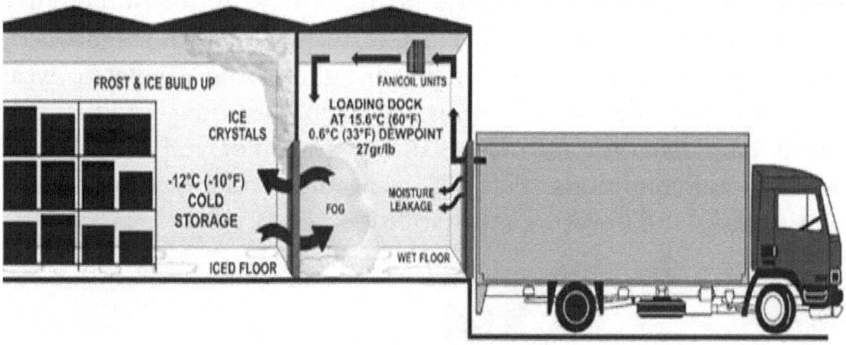

Figure 16: Market Logistics

The Indian government has embarked on an ambitious project called Sampada, a national scheme to develop an integrated supply cold chain for agriculture produce with initial investment of Rs. 6,000 crore (€764 million).

Biotechnology plays a pivotal role in developing crops with enhanced traits, tailored to withstand the challenges posed by climate change and pests. Genetic engineering techniques enable the creation of drought-tolerant varieties, capable of thriving in water-stressed environments.

Similarly, crops engineered for resistance to pests and diseases reduce the reliance on chemical pesticides, promoting eco-friendly farming practices. Biotechnological innovations also extend to the enhancement of nutritional profiles in staple crops, addressing malnutrition challenges prevalent in certain regions of India.

Figure 17: Use of IT in Agriculture

In conclusion, as India strides forward in its journey of progress and development, it is imperative to recognize the integral role of agriculture and the pivotal contributions of our farmers. From the historic achievements of the Green and White Revolutions to the present-day challenges and opportunities, agriculture remains central to our economic stability and food security. The confluence of traditional wisdom with modern technology holds the promise of a brighter future for Indian agriculture. By embracing innovations such as digitalization, precision farming, rainwater harvesting, robotics, artificial intelligence, and bio-technology, we can empower our farmers, enhance productivity, and foster sustainable practices. As we embark on this transformative journey, let us not forget the timeless values of resilience, perseverance, and unity, ensuring that every step we take towards agricultural prosperity echoes the spirit of *Jai Jawan, Jai Kisan, Jai Vigyan, Jai Anusandan.*

Sustainable Transformation of Dairying in India

Meenesh Shah[1]

Introduction

Climate change is the dmost serious environmental challenge that humanity will face in the near future. Climate Change is threatening the well-being of the current and future generations by transforming our planet's ecosystem. As per Intergovernmental Panel on Climate Change (IPCC) of the United Nations, global surface temperature change by the end of the 21st century is likely to exceed 1.5 to 2.0 degree Celsius relative to pre-industrial period. Global warming would result in long-lasting changes in all components of the climate system, increasing the likelihood of severe, pervasive and irreversible impacts for people, food security, livelihoods and planetary ecosystem. Since last few years, we are already witnessing adverse impacts of global climate change. Some of these impacts are changes in rainfall pattern, increased incidence of heat waves, droughts, floods and cyclones. These events are adversely affecting agriculture and allied sectors, and the rural people who are dependent on these sectors for their livelihood.

Agriculture and dairying are crucial for ensuring food, nutrition and livelihood securities for India. However, these sectors are highly vulnerable to risks associated with climate change.

Climate change has pronounced effect on feed production and nutrition of dairy animals. Increased environmental temperature would result in increased lignification of plant tissues which would affect feed digestibility and milk productivity. Water scarcity due to climate change would also affect feed and fodder production for animals. Climate change would increase incidence of vector-borne diseases, disease outbreaks, heat stress, reduce feed intake, milk yield and reproduction in animals. Climate change would also increase feed, water and shelter requirement of animals. If managed properly, agriculture and dairying would significantly

---

[1] Chairman, National Dairy Development Board, Anand, Gujarat, India.

contribute to mitigation of climate change and improve socio-economic sustainability.

Growth of Agriculture and Allied Sectors in Amrit Kaal

In August 2022, India completed 75 years of independence, making this year a momentous one. Coinciding with this historical milestone, the Hon'ble Prime Minister Shri Narendra Modi described the next 25 years as Amrit Kaal (leading to 2047, when India will complete 100 years of independence) and set a goal of making India a developed nation or *Viksit Bharat* by 2047.

Population growth, increasing per capita income, changing consumer preferences/consumption patterns are driving forces for increasing demand of food by middle of this century. According to United Nations (2019), India's population will increase from 1.38 billion in 2020 to 1.50 billion in 2030 and 1.59 billion in 2040. Besides meeting the increased demand of food for additional population, there is also a pressing need to increase per capita intake of food to address issues related to hunger and malnutrition and that too in environmentally sustainable way. Agriculture and allied sectors (dairying) are vital for livelihood, food and nutrition security, and socio-economic development in the country. These sectors would play a pivotal role in achieving the goal of Viksit Bharat, inclusive development, green growth and gainful employment during *Amrit Kaal.*

To achieve the goal of Viksit Bharat, agriculture and allied sectors would have to focus on sustainable use of natural resources (land and water), mitigating greenhouse gas (GHG) emissions, increasing crop and animal productivity and efficiency in cost-effective manner, linking food production with health and nutrition (addressing nutrition through healthy diets), combating climate change, improving sustainability, modernization, digitalization, natural/organic farming, renewable energy use, and significant and sustained increase in farmers' income, etc.

Climate Smart Dairying

Climate smart dairying is the need of the hour because of its immense potential to improve socio-economic and environmental sustainability of the sector, and thereby contributing to meeting the goal of making India a developed nation or Viksit Bharat by 2047. National Dairy Development Board (NDDB) has affiliation to the strong network of 1.9 lakh village dairy cooperative societies, 245 Milk Unions and 22 milk marketing federations and reach to 17 million dairy farmers. NDDB has undertaken several initiatives addressing the major aspects of dairy sustainability. These include improved agricultural practices (fodder productivity and water security), productivity enhancement though scientific breeding, nutrition and health, manure management, use of renewable energy and improving energy use efficiency. Higher dairy productivity would help produce more milk from a limited number of animals. This would lead to a more efficient use of resources such as land, feed and water, resulting in lower emissions per unit of milk produced.

Productivity Enhancement Through Scientific Breeding

It is important to increase production potential of future generations for which scientific breeding efforts are required. Over last decade and a half, infrastructure has been created to record large number of animals in the country. Systematic performance recording of important indigenous cattle and buffalo breeds, in varied agro-climatic conditions in our country has provided insightful data on breeds and breed combinations that are working well in specific condition. For the first time, large number of records for indigenous breeds in their breeding tract were collected and it was visible that in certain conditions these breeds can compete well – if not better than crossbreds.

Given the fact that stress due to climate variability and availability of feed and fodder would be increasing constraints, more emphasis would be required in promoting indigenous breeds. Selected organisations across the country are carrying out Progeny Testing and Pedigree Selection programs to produce bulls of descript indigenous breeds of cattle and buffaloes to cater to the need of the country. Constraint of lack of pedigreed data is now being overcome by genomic data. The performance data combined with DNA information is helping to identify gene combinations that can work well in specific climate.

Apart from improving milk production capacity and milk quality, the genetic improvement programmes are aiming at identifying superior bulls which can produce heifers that can withstand heat stress and disease threats, reproduce regularly, efficiently utilize feed and water resources and are profitable to the farmers.

With custom genotyping chips specifically designed for Indian cattle and buffalo population, genomic selection tools are helping to accelerate genetic progress in cattle and buffalo population covered under Artificial Insemination. Further, genotype data is helping to understand breed composition of animals that are selected and level of exotic inheritance that can work well in specific feeding and management conditions for producing better crossbreds.

NDDB is also conscious of the need to reduce the contribution of dairy animals to GHG emissions – either directly from enteric fermentation and manure or indirectly from activities related to feed and fodder production. In addition, efforts are being made to collect data on individual animal wise methane emission, milk production and feeding that would help understand methane emission per kg of milk produced in different feeding systems and take necessary steps to mitigate the same. There is sufficient genetic variability in enteric methane emission at same level of milk production among animals. This genetic variability may be exploited to select bulls that produce progenies with more milk and less methane emission per kg of milk. This will provide population wide benefits without extra cost. The combination of breeding programs aiming for higher productivity, improving feed conversion efficiency and feeding them balanced rations appears to be the practical route to reduce methane emission from the dairy sector.

Further, technologies such as sex-sorted semen, *in vitro* embryo production and embryo transfer are also being promoted. Research is being done to make these technologies affordable to the farmers to fast track genetic progress. With future animals produced by systematic breeding, that will yield more milk (with better quality) per kg of feed consumed, without being affected by increased temperature and disease threats, dairying will be more sustainable and profitable to farmers.

Enhancing Green Fodder Productivity

Fodder Seed Production of Improved Varieties

It is estimated that by year 2030, India would face shortage of green fodder by 24.6%. By years 2040 and 2050, this storage would be 20.2 and 18.4%, respectively (NITI Aayog, 2018). At present, the area under green fodder production is about 9.2 million hectares, which is almost constant over the last few years. Supply of green fodder can be enhanced significantly if quality seed of improved genetics is used to improve the fodder productivity. Seeds are a basic and critical input for agricultural production, however, at present only 25 to 30% of required quantity of fodder seed of improved varieties/hybrids is available in the country against the estimated annual requirement of 3.55 lakh metric tonnesor

MT (Sharma, 2021; Singh *et al.*, 2022). By recognizing the pivotal role of fodder seeds in achieving feed security, the Department of Animal Husbandry and Dairying (DAHD), Government of India and the NDDB have been promoting the production of quality fodder seeds. The emphasis on quality fodder seed production has been evident since Operation Flood, with NDDB spearheading organised production programmes since 1988. Notable milestones include the establishment of the Northern Region Fodder Seed Grid in 1992 and the subsequent establishment of six seed units under Operation Flood-III. Post-National Dairy Plan-I (NDP-I), significant progress has been made, with the establishment of nine seed processing units and the production of approximately 40,000 quintals of fodder seeds by dairy cooperatives.

Aimed at increasing the availability of good quality fodder seeds, the Government of India is implementing the realigned National Livestock Mission (NLM) from year 2021–22, in which NDDB is facilitating dairy cooperatives to take up fodder seed production in a big way under the aegis of this scheme. NDDB is facilitating the dairy cooperatives in project planning and approval, procurement of breeder and foundation seeds of improved varieties from Indian Council of Agricultural Research (ICAR) and Agricultural Universities, technical and financial assistance in implementation of seed production activities and linking of the seed producing agencies with those milk unions and federations which are unable to produce their own seed but require fodder seed for their farmers. Under the NLM, NDDB has facilitated dairy cooperatives in producing about 1,49,310 quintals of high-yielding fodder seed varieties from 2021–22 to 2023–24. This production progress reflects a concerted effort to bolster fodder seed production, ensuring sustainable fodder supply for economic dairy farming in the country.

Fodder Plus Farmer Producer Organisation (FPO)

Owing to the chronic shortage of dry and green fodder and its rising costs, dairy farmers are finding it difficult to make dairying as a remunerative business proposition. This shortage of quality feed and fodder is not only affecting the productivity of dairy animals but also taking away a major portion of profit from dairy farming. This situation calls for development of organised supply chain of dry and green roughages across the country in addition to concentrates.

Considering the prospect of fodder cooperatives in mitigating fodder scarcity in the country, NDDB has been designated as Implementing

Agency for formation and promotion of 100 'Fodder Plus' Farmer Producer Organisation (FPO) during FY 2022–23 with the help of Milk Unions/Producers' Organisations/NGOs/other such institutions. Fodder Plus FPO is an organisation of dairy and agricultural farmers (a group of minimum 300 farmers) for the purpose of taking up fodder development and animal husbandry activities with the objectives of enabling market access to small and marginal farmers, and providing opportunities to rural youth/fodder growers for becoming entrepreneurs and taking up fodder development/dairying activities. The Fodder Plus FPOs would engage in fodder development activities such as production and/or sale of green fodder/silage, crop residues, fodder seeds, feed supplements etc. and also take up other animal husbandry activities. It is expected that setting up of Fodder Plus FPOs would help in creating infrastructure for fodder activities and thus increasing the availability of quality fodder in the country.

Productivity Enhancement Through Scientific Nutrition

Animal Nutrition initiatives of NDDB are aimed at improving milk productivity and profitability, and reducing GHG emissions, thereby ensuring sustainable dairy production in the country.

Ration Balancing Programme

Generally, rations fed to dairy animals are nutritionally imbalanced, which results in lower milk production, less income to farmers and higher enteric methane emission per kilogram of milk. To tackle this issue, NDDB conceptualised 'Ration Balancing Programme' (RBP) and educated milk producers on scientific feeding of their animals. Under NDP-I, doorstep ration advisory service was provided for 2.8 million cattle and buffaloes belonging to 2.2 million dairy farmers in 18 states of the country (Figure 1).

As a result, there is 3% improvement in milk production (7.06 *vs.* 7.33 kg/d), 12% reduction in cost of feeding (Rs. 135.1 *vs.* 119.1 per animal per day), increase in net daily income of farmers by Rs. 27.5 per animal and 13.7% reduction in enteric methane emission (World Bank, 2020). In addition, feeding nutritionally balanced rations to animals help reduce water footprint of milk (freshwater used to produce a kg of milk) by 15% (NDDB, 2019). These findings revealed that, for every one kg of fat-and-protein corrected milk (FPCM) production, there is a saving of about 155 litres of freshwater which is mainly attributed to indirect water use for feeding (Figure 2). There is a substantial scope for reducing the

Figure 1: Ration Balancing at Farmers' Doorstep

water footprint of milk through adoption of scientific feeding practices by farmers in the country. Overall, ration balancing improves productivity of dairy animals and profitability of smallholder farmers in an environmentally sustainable manner. NDDB would continue to promote the concept through National Digital Livestock Mission (NDLM), Japan International Cooperation Agency (JICA) and upcoming programme – National Dairy Plan-II.

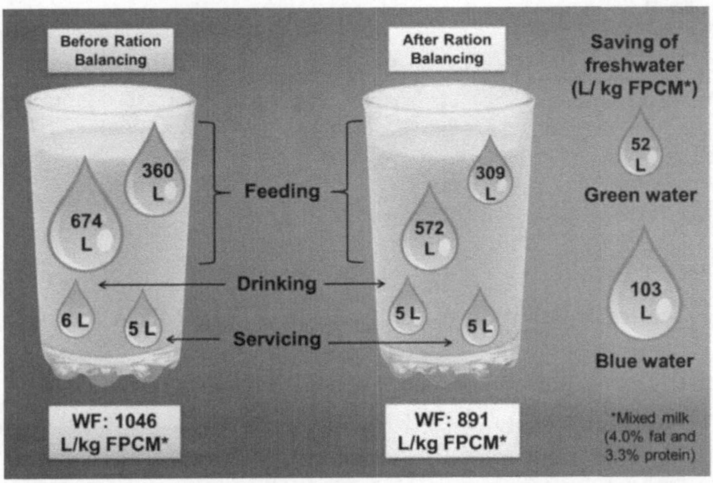

Figure 2: Effect of Feeding Balanced Rations on Water Footprint of Milk

Crop Residue Management and Total Mixed Rations

Approximately, 92 million metric tonnes (MMT) of crop residues are burnt every year in India (Meena *et al.*, 2022). Crop residue burning has negative effects on environment as well as on agricultural soils and human health. Under NDP-I, NDDB demonstrated use of several types of mowers and pick-up devices among farmers to secure crop residues from field.

Further, NDDB has introduced the concept of 'dry-Total Mixed Ration' (TMR) for feeding of dairy animals wherein, secured crop residues are incorporated in the TMR. This initiative not only help improve milk productivity but also minimise the GHG emissions associated with crop residue burning. Under the NDP-I, technical support was provided for establishment of two TMR plants. About 2000 MT of crop residues is utilized every year for production of about 3500 MT TMR blocks and pellets by these plants, thereby helping to avoid GHG emissions into the environment. TMR feeding also improved milk production of animals by 12% and avoided burning of crop residues, thereby, contributed to reduce carbon footprint of milk.

Conventional TMR: Small, marginal and landless farmers often face challenges in providing balanced rations to their animals on account of irregular supply of feed and fodders coupled with shortage of feed resources. NDDB has designed silage/green fodder, dry fodder and concentrate based **'ready-to-eat' packed TMR** (Figure 3).

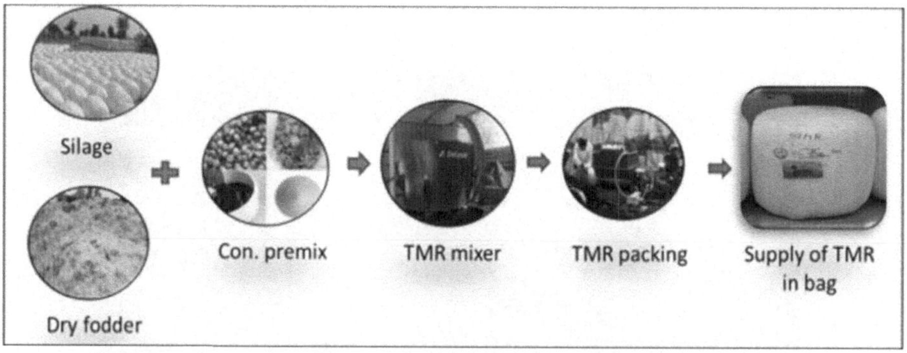

Figure 3: Ready-to-eat/conventional Total Mixed Ration

TMR feeding revealed an increase in milk yield, fat percent and net daily income by 12% (10.71 *vs.* 11.94 kg/d), 11% (3.81 *vs.* 4.24%) and Rs 38 (Rs. 104 *vs.* 142 per animal), respectively(Patel and Patel, 2022). In addition, enteric methane emission was also reduced by 11%in lactating animals. Looking at the encouraging results, it is now proposed to popularise green fodder/silage based TMR using the funding support under NLM scheme which provides 50% subsidy on the total project cost (maximum up to Rs. 50 lakh). NDDB is extending technical support to the interested milk unions for setting up of manufacturing plants for 'ready-to-eat' TMR.

Nutritional Strategy for Optimising Milk Quality

Fat and SNF are important constituents that determine the milk price. Inadequate feeding practices such as lack of sufficient energy and protein in ration, excessive concentrate and less roughage intake, negative energy and protein balance, low body condition, physiological/metabolic stress, etc. often results in lower milk fat and SNF, thereby less price realisation of milk. Keeping this in view, NDDB regularly advise dairy farmers for optimising feeding practices to enhance milk fat and SNF. Further, NDDB has also developed a feed supplement – '*Samvriddhi'* for improving milk fat and SNF content. Supplementing animals with *Samvriddhi* during early to mid-lactation period improved milk fat and SNF by 7.2 to 9.3% and 1.8 to 2.4%, respectively, in cows and 2.4 to 3.9% and 1.6 to 1.8%, respectively, in buffaloes. Increase in net daily income per animal was Rs. 14 to 25 in early and Rs. 6 to 11 in mid lactating cows and buffaloes (NDDB, 2019). With technical support of NDDB, several milk unions are taking production of *Samvriddhi,* which is benefiting dairy farmers in their milksheds.

Nutritional Strategy for Mitigation of Heat Stress

With increasing environmental temperature and humidity, dairy animals often suffer from 'heat stress'. This effect is more pronounced when temperature-humidity index (THI) exceeds a threshold of 72, mostly during summer season. As a result, feed intake and milk production of animals drops by 5 to 20% and 10 to 30%, respectively. Heat stress has negative effect on milk fat and SNF content, and also on reproductive performance of animals.

To minimise losses suffered due to heat stress, NDDB is encouraging farmers to optimise feeding practices of dairy animals during summer months. These includes changing feeding time (feeding during cooler hours), incorporating good quality concentrates in ration (replacing BIS-II cattle feed with BIS-I type feed), increasing proportion of green fodder/silage and reducing dry fodder in ration to improve digestibility and minimise heat of fermentation during digestion, chaffing of green and dry fodders, and mixing them with concentrates, feeding Total Mixed Rations, ensuring adequate availability of drinking water, supplementing bypass fat (100 to 150 g/d), using potassium rich mineral mixtures, area specific mineral mixture (100 to 150 g/d), ensuring supply of vitamins and buffers in rations of animals, etc. Apart from these, NDDB has also developed a feed supplement – *'Pashu Sheetvardhak'* to minimise the productivity loss and maintain physiological parameters of animals during summer season. Supplementation of *'Pashu Sheetvardhak'* to crossbred cows and buffaloes during the summer season resulted in additional net daily income of Rs. 20 and Rs. 36 per animal, respectively (NDDB, 2019).

Circular Bio-economy

The livestock sector in the country is already facing acute shortage of feed and fodder resources. At the same time, there is a huge opportunity to bridge the gap between demand and availability of feed and fodder for animals, if locally available agro-industrial by-products are identified and utilised efficiently for feeding animals. Such approach would help minimise wastage and convert inedible agriculture biomass into edible protein output for humans – a way towards sustainable agriculture and circular bio-economy. NDDB has already pilot tested conversion of green paddy stubble and empty pea pods into silage for feeding animals.

Conserving Green Paddy Stubbles into Silage to Mitigate Crop Residue Burning

Despite the shortage of dry fodder in many parts of the country, huge quantity of crop residue is burnt every year, with Punjab alone contributing 25 MMT. Crop residue burning not only affects the properties of agricultural soils but also results in significant nutrient loss, in addition to emitting GHGs in the environment. The negative effects of crop residue burning on the environment, soil, and human health can be avoided if this biomass is efficiently secured just after harvesting and conserved in the form of green paddy stubble silage. To develop silage-making technology for green paddy crop residue, NDDB conducted a series of trials in the laboratory and under field conditions. The trial results indicate that good-quality silage from green paddy stubbles can be produced using enzymes and silage culture.

With the encouraging trial results, NDDB has embarked on a large-scale pilot project in collaboration with MILKFED, Punjab and GADVASU,

Ludhiana for the production of green paddy stubble silage and its feeding to dairy animals. About 286 MT silage was produced by securing green paddy stubbles in Punjab. Additionally, NDDB in collaboration with GADVASU is also exploring how paddy stubble silage can be fed to buffaloes in the region. This initiative would not only help address environmental concerns related to stubble burning but also offer potential benefits to thousands of dairy farmers through augmentation of feed resources in the country.

Converting Waste from Fruit and Vegetable Processing Industries (Empty Pea Pods-EPP) into Silage

To enhance fodder resources in the country, it is imperative to explore unconventional fodder resources alongside traditional fodder crops. Fruit and vegetable wastes, including pea wastes, hold significant potential as a source of quality fodder for dairy animals. However, the challenge lies in conserving these high-moisture materials. NDDB, in collaboration with various stakeholders, is exploring the possibilities of utilising such materials for silage making. NDDB has successfully standardised the process of making silage using high-moisture pea wastes. During 2023–24, demonstration of EPP silage production was conducted in collaboration with Mother Dairy Fruit and Vegetable Pvt. Ltd. (MDFVPL) and Jharkhand Milk Federation (JMF). About 135 MT of EPP silage was produced at two different locations. The production of EPP silage would pave the way in utilising vegetable wastes, thus offering quality roughage for dairy animals at reasonable cost.

Development of Feed Additives for Methane Mitigation

Feed additives or dietary methane inhibitors have potential to reduce the methane emission between 5 to 35%. Worldwide, several feed additives are being used for mitigation of enteric methane mitigation. However, none of them have dual benefits, i.e. improvement in milk productivity as well as methane mitigation. In India, dairy farmers require feed additives that can improve milk productivity as well as reduce methane emission. NDDB is already working in this direction. Once developed, such feed additives would be recommended to Cattle Feed Plants (CFPs) for inclusion in cattle feed formulations or to farmers for direct supplementation to animals. This would help achieve large scale methane mitigation as well as further contribute to improve milk productivity/milk production efficiency of animals in the country.

Productivity Enhancement Through Animal Health

To improve the productivity of dairy animals, proper animal health services which include preventive and curative treatment, needs to be provided at the doorstep of dairy farmers, in a cost effective and efficacious manner, apart from genetic improvement and balanced nutrition programmes.

Propagation of Ethno Veterinary Concept

Ethno-veterinary Medicine (EVM) is a traditional practice of herbal veterinary medicine, which has been used for ages in India to manage ailments in livestock. It provides a simple, cost-effective and efficacious option to the dairy farmers, who are mainly marginal or landless and can ill afford costly treatment. It also provides a prompt management option to those who are out of reach of the veterinary delivery system in remote areas. The potential of EVM in providing safe milk at minimal cost is immense as it plays an important role in reducing drug usage, especially antibiotics, which would in turn help in stalling the emergence of antimicrobial resistance (AMR) – a major emerging public health concern. It also reduces the treatment costs drastically since most of the ingredients used in EVM preparations are available in the farmers' homestead.

NDDB is propagating the concept of EVM through a project – 'Disease Control through Alternative Methods' under which 15 milk unions and producer organisations across India are implementing the project with technical and financial support from NDDB. The main aim of the project is to propagate the EVM concept amongst the livestock owners and to reduce the antibiotic usage in disease treatment. As of March 2023, more than 8.3 lakh cases of EVM interventions have been documented from the project regions with an overall cure rate of 80% (NDDB, 2023). Milk unions that have been propagating the EVM concept extensively have been able to reduce their antibiotic usage by 60% over the past 4 years since implementation in 2017–18, indicating that the farmers are now increasingly opting for EVM to treat common ailments in their animals (Figure 4), a testimony to the success of knowledge transfer. Surveillance of mastitis pathogens of zoonotic importance is also being carried out in animals and humans in the project regions to study its implication and to help develop a sustainable One Health control model for diseases such as mastitis.

Figure 4: Mastitis Control Through Ethno-veterinary Medicine

Brucellosis One Health Control Programme

With aim to reduce the prevalence of brucellosis both in humans and animals, NDDB is implementing a One Health model of brucellosis control since 2012–13 in Gujarat. Recognising the significance of this zoonosis in humans, which often goes underdiagnosed and hampers productivity of both the animals and farmers, the model aims to understand the link between disease in animals and humans. So far, 3,226 farmers and animal health workers have been tested and all the 131 patients with brucellosis symptoms who received treatment have recovered fully (NDDB, 2023). Under the 'Brucellosis One Health control programme', around 5.27 lakh calf-hood vaccinations have been carried out in the project area (Figure 5). As a result, the sero-prevalence of brucellosis has been reduced from 14.1 to 8.7% in Anand and from 13 to 4.1% in Kutch district in last 5 years (Kumar *et al.* 2023).

Figure 5: Brucellosis Prevention Through Vaccination

Manure Management Initiative

With the continuous efforts in Manure Management domain since year 2018, NDDB undertook various innovations for the efficient management of the dung (Figure 6). These initiatives have now resulted into development of different Manure Management Models wherein the core focus is not only in the efficient production and use of clean energy but also in propagation of sustainable agricultural practices.

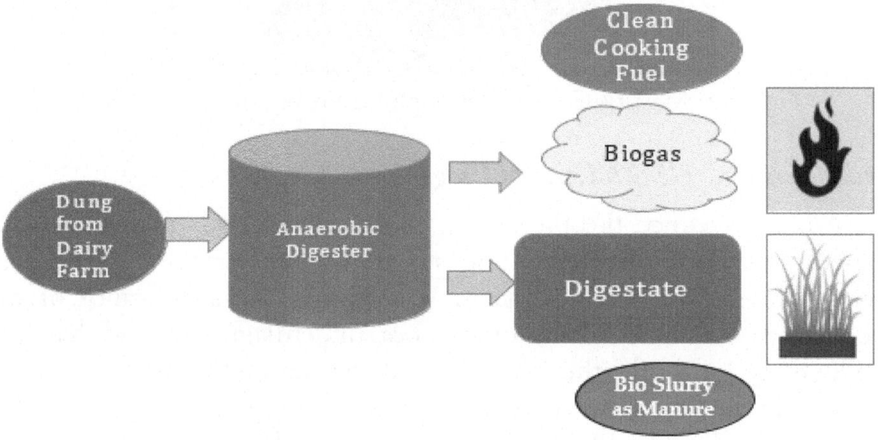

Figure 6: Manure Value Chain Model

Decentralised—Zakariyapura Model

To capitalise the abundant availability of dung in the rural area and making the rural households self-sustainable (Aatma Nirbhar) by sufficing the requirements of energy (fuel for kitchen) and fertiliser, NDDB started the manure management initiative with few farmers in one village of Anand district in Gujarat which subsequently gained prominence as 'Zakariyapura Model'.

Under the model, flexi biogas plants of 2 cubic meter size were installed in the backyard of the dairy farmers with the initial investment of Rs. 25,000 per plant (Figure 7). With the biogas plant, the beneficiary can produce clean energy equivalent to 1.5 to 2 LPG cylinders per month costing around Rs. 1,500 to 2,000. Moreover, the slurry produced from the biogas plant (having higher nutrients than Farm Yard Manure) is being utilised in the farm as a manure which is helping to reduce the dependency on the costly chemical fertilisers and improving the soil fertility.

Figure 7: Flexi Biogas Plant

To establish the end-to-end manure value chain, India's first all women manure cooperative is established to process the surplus slurry at central slurry processing centre (within the cluster of biogas plants), which converts the slurry into slurry-based value-added organic fertilisers. Hence, with the sale of the surplus slurry to manure cooperative, each beneficiary earns Rs. 1,000 to 2,000 per month depending on the quality and quantity of slurry sold.

Put together with the setup of end-to-end manure value chain, beneficiaries save or earn Rs. 3,000 to 4,000 per month which not only recovers the initial investment in the first year itself but also improves the overall wellbeing by addressing the Sustainable Development Goals such as No. 1 (No Poverty), 3 (Good Health and Wellbeing), 5 (Gender Equality), 7 (Affordable and Clean Energy), 12 (Responsible Production and Consumption) and 13 (Climate Action).

Under Manure Management initiative, more than 25,000 household level biogas plants have been installed till date with support from NDDB under various schemes of Government of India/Corporate Social Responsibility support by companies/NDDB's own funds, etc. Moreover, on the basis of success of the Zakariyapura model, several manure management initiatives at the community/cluster level with biogas units at individual household level and centralised processing have been taken up in 11 locations (9 states) across the country. Further, NDDB's Zakariyapura model is now included in the GOBARDhan scheme of Government of India, with NDDB as a technical partner under the scheme.

Centralised—Varanasi Model

With financial and technical support from NDDB, Varanasi Milk Union has setup a cow dung-based biogas plant which has a capacity to generate 4,000 cubic meter biogas daily. This is one of its kind of centralised model in which cow dung is procured from the farmers and the biogas generated is utilized to meet the thermal and electrical energy needs of the dairy plant for processing of milk and generation of steam etc. This model is known as 'Varanasi Model' (Figure 8).

At the 100% capacity, the plant needs 100 MT of dung per day which is aggregated from farmers and Gaushalas within radius of about 10 to 15 km of the dairy plant to produce biogas every day. The bio slurry produced from the biogas plant is separated into solid and liquid fractions to manufacture Phosphate Rich Organic Manure (PROM). This not only promotes use of green energy but also reduces the processing cost.

Figure 8: Centralised Varanasi Model

While on the one hand farmers are getting price of cow dung along with milk, on the other hand along with meeting the energy requirements of the dairy plant, organic fertiliser is being produced which is improving the agricultural productivity and also improving the soil health. The plant was inaugurated by Hon'ble Prime Minister Shri Narendra Modi on 7thJuly 2023. With the use of the biogas in the dairy plant, it replaced the use of Light Diesel Oil (LDO) and hence, the processing cost of milk reduced by Rs. 0.40 to 0.50 per litre. Based on the learning, NDDB is in process to take forward the 'Varanasi Model' to various locations such as Sabarkantha Milk Union in Gujarat and Barauni Dairy in Bihar.

Centralised—Banas Model

As a part of our Hon'ble Prime Minister's vision to increase farmers' income, a ground-breaking initiative in the Manure Management domain was taken by Banas Dairy by setting up of large-scale biogas unit with the raw biogas generation capacity of 2,000 cubic meter in the year 2020. The pilot project was setup in the Dama village of Banaskantha district of Gujarat which has the capacity to generate Bio-CNG from cow dung. It yields approximately 2,000 cubic meters of raw biogas daily. The generated raw biogas is compressed and purified before using as Bio-CNG for the vehicles and the bio-slurry is being utilised for the production of organic fertilisers which is now known as 'Banas Model' (Figure 9).

The primary raw material used in the biogas production process is the cow dung. The plant engages in procuring 40 MT of dung on daily basis from a network of about 250 dairy farmers situated in the nearby 12 villages. The procurement process is executed through a well-organised system involving dedicated routes in line of milk collection system. The dairy farmers are paid an average of Rs. 1.0 per kg of dung supplied. The digested slurry is further processed to produce solid fertilisers such as PROM, Fermented Organic Manure (FOM) and liquid fertilisers. Overall, the Banas model helps provide additional income to farmers, convert cow dung to wealth, improve hygiene and environment, produce nutrient rich fertiliser and generate clean energy.

Suzuki R&D Centre India Pvt Ltd (SRDI) – a fully owned subsidiary of Suzuki Motor Corporation (SMC), Japan approached NDDB to explore possibilities of collaboration for promoting dung based Compressed Biogas (CBG) as a vehicle fuel. In this context, NDDB, SRDI and Banas Dairy have signed an agreement to take up four such CBG projects in Banaskantha district to replicate the 'Banas Model'.

The major objective of this collaboration is to design, develop, implement and scale up innovative business models to efficiently utilise cow dung as a source of energy for fuelling transportation needs and as a rich source of organic fertiliser while achieving carbon neutrality. SRDI wishes to promote dung based CBG stations across the country with the help of

NDDB, so that CBG based affordable rural transportation can be promoted which would in turn help in achieving carbon neutrality.

Figure 9: Centralised Banas Model

Renewable Energy Generation

Sustainable use of groundwater is crucial for future agricultural security. By recognising this, NDDB has successfully demonstrated a pilot on how farmers can sustainably use groundwater for irrigation through solar energy. Using solar pumps, required groundwater can be extracted for irrigation and surplus electricity can be transferred to the grid for generating additional revenue for farmers.

Under this novel initiative, a solar pump owners' cooperative was formed wherein 11 farmers from Mujkuva village (Anand, Gujarat) have installed solar pumps (150 kWp capacity) for irrigating their own fields and established their own micro-grid to sell excess energy generated from the solar panels to Madhya Gujarat Vij Company Ltd. or MGVCL (Figure 10). These farmers have foregone their claims on highly subsidised agricultural electricity connection for irrigation. Mujkuva Solar Pump Irrigators Cooperative Enterprise (MSPICE) or Mujkuva Saur Urja Utpadak Sahakari Mandli was inaugurated by Hon'ble Prime Minister of India on 30th September, 2018.

This model is helping in conservation of groundwater and optimising energy consumption as the farmers use the irrigation pumps judiciously to maximise their income from sale of solar energy to the grid (about Rs. 3,000 per month). Apart from becoming 'Aatma Nirbhar' in meeting energy needs for irrigation and reducing carbon emissions up to 1.1 lakh kg Carbon di-oxide equivalent or CO_2-e annually (MSPICE, 2022), the MSPICE farmers are also saving Government subsidy on electricity (about Rs. 6.5 lakh per year). The MSPICE model has been a reference for Suryashakti Kisan Yojana (SKY), Government of Gujarat scheme for promoting feeder level grid connected solar pumps and for designing component C of Kisan Urja Suraksha evam Utthan Mahabhiyan (PM-KUSUM) scheme of Government of India for grid connected solar pumps.

Figure 10: Mujkuva SPICE Model

Organic Farming

Regenerative agricultural practices/organic farming is a key to future agricultural security and sustainability. NDDB is the chief promoter of National Cooperative Organics Limited (NCOL). The organisation would offer institutional assistance in consolidation, certification, testing, procurement, storage, processing, branding, labelling, packaging, and providing logistic facilities for promoting organic farming in the country.

The support would also be extended to marketing organic products and arranging financial aid for organic farmers through member cooperatives, including Primary Agricultural Cooperative Credit Societies (PACS)/ FPOs. Additionally, the organisation would undertake all promotional and developmental activities related to organic products in collaboration with various Government schemes and agencies. If more farmers can be convinced to practice organic farming/adapt regenerative agricultural practices through support for marketing of their products, it would help in future agricultural security and sustainability.

Green/Carbon Credits

To scale up sustainability initiatives across the county, NDDB has submitted various policy level inputs especially for suitable funding support for propagation of renewable energy applications and scaling up productivity enhancement measures.

NDDB along with Sustain Plus Energy Foundation (a Tata Trusts initiative) is working for the generation of carbon credits from the installation of the flexi biogas plants across 9 locations in 7 states in India. A mechanism would be developed wherein the generated carbon credits can be used to incentivise the farmers by providing additional avenue of earning so as to ensure regular up-keep of the biogas digesters. The project has been registered under the Voluntary Carbon Standard (VCS) mechanism.

Under the "Gobar se Samruddhi", NDDB Mrida Ltd., a wholly owned subsidiary of NDDB has partnered with the biogas plant manufacturer Sistema.bio to harness upfront carbon financing potential for installation of flexi biogas plants. The programme is being implemented through Dairy Cooperatives and voluntary agencies. With the carbon financing model, it has helped to offer 2 cubic meter capacity biogas plants at just Rs. 6,000 against the cost of Rs. 35,000.

Government of India's Green Credit Program would provide further incentives for adoption of sustainability measures in dairying. The notification on Green Credit Program has identified eight sectors. In most of these, dairy sector has already taken several initiatives. The milk unions can earn green credits for their initiatives in manure management, afforestation drives, treatment of dairy wastewater, use of renewable energy in dairy value chain, and efficient design resulting in reduction in material and energy intensity in dairy processing.

Towards Net-Zero Dairying

By recognising threats associated with climate change, majority of countries across the world adopted the Paris Agreement with an aim to pursue efforts to limit global warming to 1.5 degree Celsius above pre-industrial levels. To achieve this goal, world leaders have agreed to take intense climate change mitigation actions in a move towards Net-Zero GHG emission targets by 2050 during COP26 in 2021. India also committed to achieve Net-Zero target by 2070.

NDDB is promoting sustainable practices for mitigation of GHG emissions and improving socio-economic sustainability of the dairy sector in India. NDDB is encouraging dairy farmers in the country to adopt best available scientific farm management practices and also to dairy processing units to take up sustainable initiatives. Enteric fermentation is a major hotspot contributing about 70% to total cradle to farm-gate GHGs. Manure management, feed production and on-farm energy use contribute about 15, 10 and 5%, respectively. Adoption of sustainable practices such as scientific fodder production, breeding, feeding, health, manure and energy management coupled with GHG removal through Carbon sequestration would help achieve the state of 'Net-Zero emissions' by 2070. A framework of key initiatives for Net-Zero dairying including genetic improvement, maintaining animal health, scientific fodder and feeding management, manure and energy management and Carbon sequestration is being promoted by NDDB.

To create favourable environment for successful adoption of Net-Zero practices by dairy sector, farmers and milk unions in the country need to be supported with appropriate incentives and policies from the government.

Way Forward

The agriculture and dairying continue to play a vital role in the country's socio-economic development and progress. These sectors have tremendous potential to fulfil India's dream to become a developed country (Viksit Bharat) by 2047. However, to achieve this goal, consistent and long-term commitments/action plans for adoption of sustainable practices such as optimising natural resource use, mitigation of GHG emissions to combat climate change, improving productivity and efficiency in cost effective way, scientific breeding, feeding and health practices, efficient manure management, modernisation, digitalisation, natural farming/regenerative agriculture, focus on renewable energy, healthy and sustainable food production, and enhancing farmers' income, etc. would be required. Additionally, strengthening policy framework, enforcing appropriate regulations, a well-coordinated strategy between central and state governments for ensuring next revolution in agriculture and dairying, capacity building and skill development of the workforce involved in these sectors, and behavioural changes of producers, processors and consumers involved in entire supply chain would together contribute to sustainable transformation of agriculture and dairying in Amrit Kaal.

References

[1] Kumar, A.V.H., Shroff, S., Dutta, P., Dash, S., Prasad, A., Bahekar, V.S., Sarangi, L.N., Reddy, R.V.C. and Gupta, R.O. (2023). A One Health field model for control of bovine brucellosis in Indian settings. In: Proceedings of 23rd Indian Veterinary Congress and 30th Annual Conference of IAAVR & National Symposium on 'Advanced in Tackling Antimicrobial Resistance and Ensuring Food Safety under One Health Perspective', Kamdhenu University, Anand, India, pp. 42–57.

[2] Meena, H.N., Singh, S.K., Meena, M.S., Narayan R. and Bheem Sen (2022). Crop Residue: Waste or Wealth? Technical Bulletin 2022. ICAR-Agricultural Technology Application Research Institute, Zone-II, Jodhpur, pp. 1–30. Available: https://krishi.icar.gov.in/jspui/bitstream/123456789/71699/1/CROP%20RESIDUE%20WASTE%20OR%20WEALTH.pdf

[3] MSPICE (2022). Second Bi-Annual Progress Report 2020–22. Mujkuva Solar Pump Irrigators' Cooperative Enterprise (MSPICE), Anand, Gujarat.

[4] NDDB (2019). Annual Report 2018–19. National Dairy Development Board. Anand, India. Available: https://www.nddb.coop/sites/default/files/NDDB-AR-2019-ENGLISH-24022020.pdf

[5] NDDB (2023). Annual Report 2022–23. National Dairy Development Board. Anand, India. Available: https://www.nddb.coop/sites/default/files/pdfs/NDDB_AR_2023_090124_Eng.pdf

[6] NITI Aayog (2018). Demand & supply projections towards 2033. Crops. Livestock, Fisheries and Agricultural Inputs. The Working Group Report, NITI Aayog, Government of India. Available: https://www.niti.gov.in/sites/default/files/2023-02/Working-Group-Report-Demand-Supply-30-07-21.pdf

[7] Patel, A.D. and Patel, N.R. (2022). To study, plan, and introduce TMR, with necessary composition, at Virpur and Balasinor region of Kaira milk shed area. Project Work Module Segment submitted to Kaira District Cooperative Milk Producers' Union Limited, Anand. Institute of Rural Management, Anand, Gujarat.

[8] Sharma, T.R. (2021). Forage security for sustainable livestock development. 60th Foundation Day Lecture, ICAR-Indian Grassland and Fodder Research Institute. Available: https://igfri.icar.gov.in/wp-content/uploads/2022/09/60th-Foundation-Day-Lecture-delivered-by-Dr.-T.R.-Sharma-DDG-CS-ICAR-on-01-Nov.-2021.pdf

[9] Singh, D.N., Bohra J.S., Tyagi V., Singh T., Banjara T.R. and Gupta G. (2022). A review of India's fodder production status and opportunities. Grass Forage Science, 77: 1–10. DOI: https://doi.org/10.1111/gfs.12561

[10] United Nations (2019). World Population Prospects 2019, Volume I: Comprehensive Tables, ST/ESA/SER.A/426. Available: https://population.un.org/wpp/Publications/Files/WPP2019_Volume-I_Comprehensive-Tables.pdf

[11] World Bank (2020). India – National Dairy Project. Implementation and Completion Results Report, No. ICR00005044. World Bank Group, Washington D.C. Available: https://documents.worldbank.org/en/publication/documents-reports/documentdetail/220451468034784073/india-national-dairy-project

Harnessing Secondary Agricultural Products for Economic Growth and Environmental Sustainability

V.P. Chovatia[1] and P.A. Pandya[2]

Introduction

Agricultural waste poses a significant economic and environmental challenge in India, with vast amounts of residues generated annually from crop stubble, husks, and other byproducts. Mismanagement of this waste leads to air pollution, soil degradation and inefficient resource utilization. According to estimates by the Indian Council of Agricultural Research (ICAR) and the Ministry of Agriculture, these waste causes economic losses amounting to billions of dollars each year due to lost potential revenue from unused biomass, increased healthcare costs related to pollution-induced illnesses and reduced agricultural productivity. To combat the issue, initiatives are exploring waste-to-energy technologies, efficient recycling methods, and improved farming practices to unlock the economic potential of agricultural waste while mitigating its adverse environmental and public health effects.

Secondary Agricultural Products for Increasing Farmers Income

Agriculture is not just about primary crop production but also involves a wealth of secondary products that can contribute significantly to farmers' income. The utilization of these secondary products is crucial for maximizing the economic potential of agriculture, especially in regions where smallholder farmers dominate. While the potential of secondary agricultural products for income generation is substantial, challenges such as limited infrastructure, lack of technological know-how and market access barriers need to be addressed. Investments in research, capacity building, and supportive policies are essential for realizing the full

[1] Vice Chancellor, Junagadh Agricultural University, Junagadh, Gujarat, India. E-mail: vc@jau.in

[2] Assistant Professor, Junagadh Agricultural University, Junagadh, Gujarat, India. E-mail: vc@jau.in

economic benefits of these secondary products. Secondary agricultural products play a pivotal role in increasing farmers' income and promoting sustainable agricultural development. Harnessing the economic potential of these products requires concerted efforts from policymakers, researchers, and stakeholders to facilitate market linkages, promote value addition and improve resource utilization. By prioritizing the development of secondary agricultural sectors, we can empower farmers, enhance rural economies and foster agricultural sustainability.

Secondary Agricultural Products for Climate Resilience Sustainable Agriculture

The importance of various secondary products from agriculture for climate resilient and sustainable agriculture cannot be overstated. Secondary agricultural products play a crucial role in enhancing climate resilience and promoting sustainable farming practices by providing alternative income sources, improving resource efficiency, and reducing greenhouse gas emissions. Here are several ways in which secondary agricultural products contribute to climate resilience and sustainable agriculture:

1. ***Bioenergy Generation from Crop Residues:*** Utilizing crop residues like rice husk, sugarcane bagasse, and wheat straw for bioenergy production reduces the dependency on fossil fuels, thereby lowering greenhouse gas emissions and promoting renewable energy sources (Kaur and Sharma, 2020).
2. ***Carbon Sequestration and Soil Health:*** Utilization of agro-processing waste for composting and organic fertilizer production enhances soil health, improves water retention capacity, and promotes carbon sequestration, contributing to climate resilience and sustainable soil management (FAO, 2019).
3. ***Diversification of Income Streams:*** Secondary products such as herbal extracts, essential oils, and value-added processed goods provide additional income sources for farmers, reducing their vulnerability to climate-related risks and market fluctuations (Sharma *et al.*, 2019).
4. ***Reduction of Food Loss and Waste:*** Value addition through processing of agricultural products into items like jams, sully, ketchup, pickles, or dried fruits ect helps in reducing post-harvest losses and food waste, which is crucial for achieving food security and minimizing resource wastage (Minot and Ngigi, 2018).
5. ***Livestock Feed and Animal Husbandry:*** Byproducts from agriculture, such as oilcakes, bran, and husks, serve as nutritious animal feed,

supporting sustainable livestock production and reducing the environmental footprint of animal husbandry (Raju *et al.*, 2017).

6. ***Promotion of Agroforestry and Sustainable Crop Production:*** Crop residues and byproducts can be used for mulching, composting, or as organic amendments in agroforestry systems, enhancing biodiversity, conserving water, and improving overall ecosystem resilience (Kumar and Sharma, 2020).
7. ***Development of Circular Economy Models:*** Repurposing agricultural waste into biodegradable packaging materials, bio-based chemicals, or bio-based materials promotes circular economy principles, reducing waste generation and enhancing resource efficiency (Sarkar and Singh, 2021).

Types of Secondary Products

1. ***Oilseeds and Oils***: Many crops such as soybeans, sunflowers, rapeseed, and olives are cultivated primarily for their oil content. These oils serve as cooking oils, industrial lubricants, biofuels, and raw materials for cosmetics and pharmaceuticals.
2. ***Residues and By-products***: Agricultural residues like straw, husks, and stalks are valuable sources of biomass. They can be converted into biofuels (like ethanol), organic fertilizers, and animal feed.
3. ***Plant Extracts and Phytochemicals***: Various parts of plants contain bioactive compounds with medicinal properties. These include polyphenols, flavonoids, alkaloids, and essential oils. Examples include curcumin from turmeric, quercetin from onions, and caffeine from coffee beans.
4. ***Natural Fibers***: Fibrous crops like cotton, jute, hemp, and flax provide raw materials for textiles, ropes, paper, and construction materials.
5. ***Plant-Based Polymers***: Starches, cellulose, and other polymers from crops are used in bioplastics, packaging materials, and coatings.
6. ***Natural Dyes and Pigments***: Certain crops yield natural colorants used in textiles, cosmetics, and food products. Examples include indigo from indigofera plants and betalains from beetroots, Anton colour for butter and card.
7. ***Animal Feed Additives***: Many crop residues and by-products serve as nutritious feed additives for livestock, contributing to sustainable animal husbandry practices.

Utilization and Applications

- ***Food and Beverage Industry:*** Plant extracts and natural flavors are used in food products and beverages for flavoring and preservation. Oils and fats serve as cooking ingredients.
- ***Pharmaceuticals and Nutraceuticals:*** Medicinal compounds derived from plants are used in traditional medicine and modern pharmaceuticals for treating various ailments.
- ***Cosmetics***: Plant-based oils, waxes, and extracts are used in skincare, hair care, and cosmetics due to their moisturizing and antioxidant properties.
- ***Textiles and Apparel:*** Natural fibers like cotton and linen are transformed into clothing and textiles, reducing the environmental impact compared to synthetic fibers.
- ***Bioenergy and Biofuels:*** Crop residues and oils are utilized in the production of biofuels like biodiesel and ethanol, contributing to renewable energy sources.
- ***Biodegradable Materials:*** Plant-based polymers and bioplastics offer sustainable alternatives to conventional plastics, reducing plastic pollution.

Below are various crop specific examples where we can harness the secondary products from various crops to ensure the more income to the farmers along with environment sustainability to combat the climate change.

Groundnut

Groundnut (Arachishypogaea), also known as peanut, is a crop that yields various secondary products apart from its primary edible nuts. These secondary products have diverse applications in food processing, cosmetics, pharmaceuticals, and agriculture.

	Product	*Application*
1.	***Groundnut Oil*** Groundnut seeds are rich in oil content (about 45–50%) and are extracted using mechanical pressing or solvent extraction methods.	• ***Cooking Oil:*** Groundnut oil is widely used for cooking due to its high smoke point and pleasant flavor. • ***Industrial Use:*** It serves as a base oil for margarine, salad dressings, and baked goods. • ***Cosmetics:*** Groundnut oil is used in skincare products and cosmetics for its moisturizing properties. (Siddiq and Sikandar, 2018)

	Product	Application
2.	***Peanut Butter*** Peanut butter is made by grinding roasted groundnut kernels into a creamy paste.	• ***Food Industry:*** Peanut butter is a popular spread and ingredient in confectionery and baked goods. • ***Nutritional Supplements:*** It serves as a protein-rich dietary supplement. (Patel and Panchal, 2019)
3.	***Groundnut Flour*** Groundnut flour is obtained by grinding roasted or defatted groundnut kernels.	• ***Gluten-Free Products:*** Groundnut flour is used in gluten-free baking and cooking. • ***Protein Supplements:*** It is a source of protein used in nutritional supplements. (Aluyor and Oyakhire, 2019)
4.	***Peanut Hulls*** The outer shell or hull of groundnuts.	• ***Animal Feed:*** Peanut hulls are used as a source of fiber in livestock feed. • ***Mulching:*** They can be used as mulch in agriculture to improve soil health. (Li *et al.*, 2018) • It has highly calorific value using for coal preparation.
5.	Peanut Shell Charcoal Peanut shells can be carbonized to produce charcoal.	• ***Fuel:*** Peanut shell charcoal is used as a clean-burning fuel. • ***Activated Carbon:*** It is used in water purification and air filtration. (Das and Ganesh, 2017)

Castor

Castor (Ricinuscommunis) is a valuable oilseed crop known for its primary product, castor oil. In addition to castor oil, castor plants offer several secondary products with diverse applications across industries.

	Product	Application
1.	***Castor Meal*** Castor meal, also known as castor cake, is the byproduct left after extracting oil from castor seeds.	• It is used as an organic fertilizer due to its nutrient-rich composition, providing nitrogen, phosphorus, and potassium to plants. • Castor meal is utilized in organic farming as a soil conditioner to improve soil structure and fertility. (Sujatha and Reddy, 2015)
2.	***Castor Husk*** The outer covering of castor seeds, has practical uses beyond the oil extraction process.	• It can be utilized as a source of biomass for energy generation through combustion or gasification. • Castor husk is employed as a raw material for making biochar, which improves soil fertility and carbon sequestration. (Balasubramanian *et al.*, 2017)

	Product	*Application*
3.	***Castor Wax*** Castor wax, a natural vegetable wax obtained from castor oil, has various industrial and cosmetic applications.	• It is used as a base material in the production of polishes, coatings, and carbon paper. • Castor wax is employed in the cosmetic industry for making lipsticks, creams, and hair care products due to its emollient properties. (Pandey *et al.*, 2020)
4.	***Castor Biomass*** Castor biomass, including leaves and stalks, can be repurposed for various beneficial uses.	• It serves as a source of organic matter for composting, enhancing soil fertility and structure. • Castor biomass can be utilized for mulching to conserve soil moisture and suppress weed growth. (Sujatha *et al.*, 2019)

Cotton

Cotton (Gossypium spp.) is a versatile crop that produces not only fibers but also several valuable secondary products with diverse applications across industries. From oil to animal feed and biofuel, different parts of the cotton plant are utilized to extract these secondary products, contributing to sustainability and economic viability.

	Product	*Application*
1.	***Cottonseed Oil*** Cottonseed oil is extracted from the seeds of the cotton plant, which contain about 18–22% oil.	• ***Cooking Oil:*** Cottonseed oil is used for cooking due to its high smoke point and neutral flavor. • ***Industrial Use:*** It is used in the production of margarine, salad dressings, and frying fats. • ***Cosmetics:*** Cottonseed oil is used in skincare products and cosmetics for its moisturizing properties. (Sammour, 2021)
2.	***Cottonseed Meal*** Cottonseed meal is a by-product obtained after extracting oil from cottonseeds. It is rich in protein (about 40–45%).	• ***Animal Feed:*** Cottonseed meal is used as a high-protein feed ingredient for livestock and poultry. • ***Organic Fertilizer:*** It is also used as an organic fertilizer due to its nutrient content. (McDonald *et al.*, 2021)

3.	***Cotton Linters*** Linters are short fibers that adhere to cottonseeds after ginning.	• ***Cellulose Products:*** Linters are used in the production of cellulose-based products such as cellulose acetate and nitrocellulose. • ***Cosmetics:*** Linters are used in cosmetic products like facial wipes and cotton pads. (Zeronian and Collins, 2001)
4.	***Cotton Stalks and Residues*** Stalks and other plant residues from harvested cotton plants.	• ***Biomass Energy:*** Cotton stalks can be used as biomass for energy production and biofuel. • ***Paper and Pulp:*** Residues can be used in papermaking and pulp production. (Filho, 2019)

Paddy

Paddy, or rice paddy, is a crucial crop that not only provides rice grains as a staple food but also yields several valuable secondary products that have diverse in various industries. These secondary products, derived from different parts of the rice plant, contribute to sustainability, waste reduction, and economic viability.

	Product	*Application*
1.	***Rice Bran*** Rice bran is the outer layer of the rice grain, separated during the milling process.	• ***Edible Oil:*** Rice bran oil is extracted from rice bran and is used for cooking and as a healthy alternative to other cooking oils. • ***Animal Feed:*** Rice bran is a valuable ingredient in animal feed due to its high nutrient content. • ***Nutraceuticals:*** Rice bran contains bioactive compounds and antioxidants used in dietary supplements. (Kong *et al.*, 2020)
2.	***Rice Husk*** Rice husk is the outer layer of the rice grain, separated during milling.	• ***Biomass Energy:*** Rice husk is used as a renewable biomass fuel for electricity generation and heat production. Also for white coal preparation.

	Product	Application
		• ***Building Materials:*** Rice husk ash is used in the production of eco-friendly building materials like bricks and insulation. • ***Animal Bedding:*** Rice husk can be used as bedding material for livestock due to its absorbent properties. (Ahmed *et al.* 2018).
3.	***Rice Straw*** Rice straw is the stalks left after harvesting rice grains.	• ***Animal Feed:*** Rice straw is used as fodder for livestock, especially cattle and sheep. • ***Mulching and Soil Amendment:*** It can be used as mulch to improve soil health and retain moisture. • ***Bioenergy Production:*** Rice straw can be used for biogas production through anaerobic digestion. (Singh and Sidhu, 2019).
4.	***Rice Water*** Rice water is the starchy water left after washing rice.	• ***Skincare:*** Rice water is used in skincare products for its soothing and brightening properties. • ***Hair Care:*** It is used in hair care products to strengthen hair and promote growth (Ariefdjohan *et al.* 2018).

Wheat

Wheat (Triticum spp.) is a staple crop that yields various secondary products beyond its primary use as a source of flour and grain. These secondary products have diverse in food processing, animal feed, biofuel production, and more.

	Product	Application
1.	***Wheat Bran*** Wheat bran is the outer layer of the wheat kernel, separated during milling.	• ***Dietary Fiber:*** Wheat bran is rich in dietary fiber and is used in food products to enhance fiber content. • ***Animal Feed:*** It is a valuable ingredient in livestock and poultry feed due to its nutritional value. • ***Health Supplements:*** Wheat bran is used in dietary supplements for its health benefits. (Schoenlechner and Berghofer,2020)

	Product	Application
2.	***Wheat Germ Oil*** Wheat germ oil is extracted from the germ (embryo) of wheat kernels.	• ***Nutritional Supplements:*** Wheat germ oil is rich in vitamin E and is used as a dietary supplement. • ***Cosmetics:*** It is used in skincare products for its moisturizing and antioxidant properties. (Sohaib *et al.* 2019)
3.	***Wheat Straw*** Wheat straw is the stalks left after harvesting wheat grains.	• ***Animal Bedding:*** Wheat straw is commonly used as bedding material for livestock. • ***Biofuel Production:*** It can be converted into biofuels like ethanol through fermentation. • ***Mulching:*** It can be used as mulch in agriculture to improve soil health. (Baiand Sun, 2017)
4.	***Wheat Gluten*** Wheat gluten is the protein component extracted from wheat flour.	• ***Food Industry:*** Wheat gluten is used as a binder and texturizer in food products like bread, pasta, and vegetarian meat substitutes. • ***Non-Food:*** It is used in non-food products like adhesives and coatings. (Janek and Jeliński, 2020)

Pigeonpea

Pigeonpea (Cajanuscajan) is a versatile legume crop that offers several secondary products with diverse applications across various industries.

	Product	Application
1.	***Pigeonpea Husk*** Pigeonpea husk, the outer covering of the seeds, is a valuable byproduct with several applications.	• It can be used as a source of biomass for energy production through combustion or gasification. • Pigeonpea husk is utilized in the production of animal feed due to its fiber content. (Hiremath *et al.* 2019)
2.	***Pigeonpea Leaves*** Pigeonpea leaves are rich in nutrients and have several practical uses.	• They can be used as green manure to enhance soil fertility and organic matter content.

	Product	Application
		• Pigeonpea leaves are utilized in livestock feeding as a source of protein and minerals. (Kumar *et al.* 2020)
3.	***Pigeonpea Stalks*** Stalks from pigeonpea plants can be repurposed for various beneficial uses.	• They can serve as a source of fiber for making handicrafts and traditional products. • Pigeonpea stalks are utilized as a raw material for making compost and organic fertilizers. (Kumar *et al.* 2019)
4.	***Pigeonpea Seeds*** Besides being a primary product, pigeonpea seeds offer secondary products with diverse applications.	• Pigeonpea seeds are processed to extract oil, which can be used for cooking and industrial purposes. • They are utilized for making pigeonpea flour, which is used in culinary preparations and as a protein-rich supplement. (Gogoi *et al.* 2020)

Banana

Bananas (Musa spp.) are not only popular fruits but also yield several valuable secondary products that have diverse in various industries, including food, agriculture, cosmetics, and textiles. From peels to fibers and extracts, different parts of the banana plant are utilized to extract these secondary products, contributing to sustainability and economic viability.

	Product	Application
1.	***Banana Fiber*** Banana fibers are extracted from the pseudostems of banana plants.	• ***Textiles:*** Banana fibers are used to make textiles and fabrics known as banana silk or banana linen. • ***Handicrafts:*** They are used in the production of handicraft items like bags, mats, and hats. • ***Papermaking:*** Banana fibers are used in the production of specialty papers. (Jahan *et al.* 2018).

	Product	Application
2.	***Banana Peel Extracts*** Extracts are obtained from banana peels, which are rich in bioactive compounds.	• ***Cosmetics:*** Banana peel extracts are used in skincare products for their antioxidant and anti-inflammatory properties. • ***Food Additives:*** They are used as natural additives in food products for their health benefits. (Prashanth *et al.* 2018).
3.	***Banana Stems and Leaves***	• ***Animal Feed:*** Banana stems and leaves are used as fodder for livestock. • ***Compost:*** They are used in composting to enrich soil fertility. (Othman *et al.* 2018).
4.	***Banana Peel Biogas*** Banana peels can be used to produce biogas through anaerobic digestion.	• ***Renewable Energy:*** Biogas produced from banana peels can be used for cooking and electricity generation. (Rajinikanth *et al.* 2018).

Sugarcane

Sugarcane (Saccharumofficinarum) is a versatile crop that yields not only sugar but also several valuable secondary products with diverse applications across various industries.

	Product	Application
1.	***Bagasse*** Bagasse, the fibrous residue left after sugarcane stalks are crushed to extract juice, has various applications. It is used as a biofuel for heat and electricity generation.	• Bagasse can be transformed into pulp for paper and board production. • It serves as a raw material for manufacturing biodegradable packaging materials. (Goyal *et al.* 2017)
2.	***Molasses*** Molasses is a byproduct of sugarcane processing and has several industrial applications.	• It is used as a feedstock for ethanol production. • Molasses serves as a component in the production of rum and other distilled spirits. • It is used as an animal feed additive due to its high energy content. (Aditi and Radhika, 2019).

	Product	*Application*
3.	***Ethanol*** Sugarcane is a major feedstock for ethanol production.	• Ethanol is used as a biofuel for blending with gasoline (e.g., E10, E85). • It serves as an industrial solvent and raw material for chemical synthesis. • Ethanol production contributes to reducing greenhouse gas emissions. (Dweikat *et al.*2020).
4.	***Sugarcane Wax*** Sugarcane wax is extracted from sugarcane stems and has diverse applications.	• It is used in the production of candles, polishes, and coatings. • Sugarcane wax serves as a raw material for cosmetic and pharmaceutical products. (Freire *et al.*, 2020)

Conclusion

From above examples, it is evident that significance of secondary agricultural products goes beyond economic advantages to include climate resilience and sustainable agriculture. These products contribute to bioenergy production, carbon sequestration, income diversification, food loss reduction, and enhancement of livestock feed. Leveraging these products can improve climate-smart practices, adaptive capacity, and overall agricultural sustainability. By harnessing secondary products from agricultural crops, we can enhance the sustainability and economic viability of agriculture while contributing to the development of eco-friendly industries. This approach aligns with the principles of circular economy and resource efficiency, making agriculture a key player in the transition towards a more sustainable future.

References

[1] Aditi, G. and Radhika, P. (2019). Molasses as a Feed Supplement: A Review. *International Journal of Livestock Research*, 9(1), 1–10.

[2] Ahmed, R., Khan, M.M.R., Zafar, A., Gupta, M.K. and Anwar, S. (2018). Rice Husk Ash: Sustainable Applications in Building and Environmental Engineering. In *Sustainable Construction and Building Materials* (First Edition), 335–350. https://doi.org/10.1016/B978-0-12-819626-6.00014-8

[3] Aluyor, E.O. and Oyakhire, O.M. (2009). Peanut and its By-products: A Review. *International Journal of Food Science & Technology*, 44(5), 441–456. https://doi.org/10.1111/j.1365-2621.2008.01837.x

[4] Ariefdjohan, M.W., Nwafor, I.C. and Mira-Salama, D. (2018). The Role of Rice Water in Hair and Skin Care: A Review. *Journal of Cosmetic Dermatology*, 17(4), 592–595. https://doi.org/10.1111/jocd.12545

[5] Bai, Y. and Sun, Y. (2017). Wheat Straw Utilization for Biofuel Production. In *Comprehensive Energy Systems* (First Edition), 140–156. https://doi.org/10.1016/B978-0-12-809597-3.00073-5

[6] Balasubramanian, V., Maheswari, C.U., Meenakshi, S. and Sundararajan, T. (2017). Potential of Castor Stalks and Husks for Bioenergy. *Journal of Renewable Energy*, 2017.

[7] Das, P. and Ganesh, A. (2017). Potential Utilization of Peanut Shell for Carbon and Energy Production. *Biomass Conversion and Biorefinery*, 7(4), 453–467. https://doi.org/10.1007/s13399-017-0263-7

[8] Dweikat, I., Awad, D. and Singh, P. (2020). Ethanol Production from Sugarcane: A Review on Feasibility and Challenges. *Journal of the Science of Food and Agriculture*, 100(8), 3208–3217.

[9] Food and Agriculture Organization (FAO). (2019). Agro-industrial Wastes for Bioenergy Generation: Potential and Challenges. *FAO Renewable Energy Assessment Programme.*

[10] Filho, G.L. (2019). Cotton Residue: Properties and Utilization. In *Sustainable Fibres and Textiles* (First Edition), 159–172. https://doi.org/10.1016/B978-0-12-813920-5.00010-0

[11] Food and Agriculture Organization (FAO). (2019). Agro-industrial Wastes for Bioenergy Generation: Potential and Challenges. *FAO Renewable Energy Assessment Programme.*

[12] Freire, C.S.R., Silvestre, A.J.D. and Neto, C.P. (2020). Sugarcane Wax as a Renewable Raw Material in the Development of Eco-friendly Composites. *Green Chemistry*, 22(19), 6450–6461.

[13] Goyal, P., Chauhan, Y.S., Singh, P. and Dhiman, R. (2017). Bagasse—A Potential Raw Material for Papermaking. *Materials Today: Proceedings*, 4(2), 1243–1247.

[14] Hiremath, G.M., Murthy, H.N.K., Sharanagouda, H. and Gangadhar, M. (2018). Nutritional Evaluation of Pigeonpea Husk on Goat Rations. *International Journal of Livestock Research*, 8(4), 195–200.

[15] Jahan, M.S., Yusoff, M.M. and Yahaya, A.H. (2018). Banana Fiber: A Review on Extraction, Properties, and Applications. *Journal of Natural Fibers*, 15(1), 1–15. https://doi.org/10.1080/15440478.2016.1265813

[16] Janek, K. and Jeliński, T. (2020). Wheat Gluten as a Raw Material for the Food Industry and Other Sectors of the Economy. In *Wheat Quality for Improving Processing and Human Health* (First Edition), 259–278. https://doi.org/10.1016/B978-0-12-816360-3.00012-9

[17] Kaur, A. and Sharma, A. (2020). Prospects of Bioenergy Generation from Crop Residues: A Review. *Renewable and Sustainable Energy Reviews*, 133, 110308.

[18] Kong, F., Wang, X. and Guo, C. (2020). Rice Bran Oil: Extraction, Composition, and Applications. In *Handbook of Food Bioengineering* (First Edition), 907–921. https://doi.org/10.1016/B978-0-12-818766-0.00049-1

[19] Kumar, A. and Sharma, S. (2020). Natural Fibers: A Sustainable Resource for the Textile and Apparel Industry. *Journal of Textile Science & Engineering*, 10(5), 1–7.

[20] Kumar, N., Bora, B.C. and Sarmah, B.K. (2020). Nutritive Evaluation of Pigeon Pea (Cajanuscajan) Foliage as Livestock Feed. *International Journal of Current Microbiology and Applied Sciences*, 9(8), 3589–3596.

[21] Kumar, P., Singh, K. and Singh, A. (2019). Assessment of Biomass, Nutrient Content, and Quality Traits of Pigeonpea (Cajanuscajan) Stover. *International Journal of Current Microbiology and Applied Sciences*, 8(1), 3067–3074.

[22] Li, S., Zhang, L. and Wang, Y. (2018). Potential Uses of Peanut By-products in Food Processing: A Review. *Journal of Food Science and Technology*, 55(8), 2837–2844. https://doi.org/10.1007/s13197-018-3263-0.

[23] McDonald, P., Edwards, R.A., Greenhalgh, JF.D. and Morgan, C.A. (2011). Cottonseed Meal and Other Oilseed Meals. In *Animal Nutrition* (Seventh Edition), 308–317. https://doi.org/10.1016/B978-0-7020-3422-5.00016-6

[24] Minot, N. and Ngigi, M. (2018). Value Addition and Processing of Agricultural Products for Enhanced Food Security and Income Generation in Kenya. *IFPRI Discussion Paper* 1747.

[25] Narwal, S.S., Tewari, L., Singh, R. and Kamal, R. (2019). Secondary Products from Crops and Agroindustrial Residues: Recent Trends in Valorization for the Food Industry. *Critical Reviews in Food Science and Nutrition*, 59(22), 3619–3647.

[26] Othman, N., Abidin, N.Z. and Jusoh, M. (2018). Potential of Banana By-Products: An Overview. *Journal of Food Science and Technology*, 55(8), 3231–3242. https://doi.org/10.1007/s13197-018-3256-z

[27] Pandey, A., Negi, Y.S. and Negi, P.S. (2020). Castor Oil: A Promising Industrial Source of Ricinoleic Acid-Based Products. *Industrial Crops and Products*, 158, 112999.

[28] Pandey, K.B. and Rizvi, S.I. (2009). Plant polyphenols as dietary antioxidants in human health and disease. *Oxidative Medicine and Cellular Longevity*, 2(5), 270–278.

[29] Patel, J. and Panchal, A. (2019). Peanuts and Peanut Butter: Processing Technology and Product Development. In *Food Processing for Increased Quality and Consumption* (First Edition), 135–153. https://doi.org/10.1016/B978-0-12-814164-7.00008-8

[30] Prashanth, K.V.H., Rajashekar Reddy, K. and Varadaraj, M.C. (2018). Potential Applications of Banana Peels as a Source of Cosmetic and Nutraceuticals Materials: A Review. *Journal of Cosmetic Dermatology*, 17(6), 907–912. https://doi.org/10.1111/jocd.12634

[31] Rajinikanth, R., Raj, M.S.A., Balaji, K. and Ramakrishnan, K. (2018). Banana Waste Biomass: A Potential Feedstock for Biogas Production.

Renewable and Sustainable Energy Reviews, 82(Part 1), 1152–1160. https://doi.org/10.1016/j.rser.2017.10.060

[32] Raju, V.R., Meenakshi, C., Naveena, K. and Radhakrishna, R. (2017). Role of Byproducts in Enhancing Livestock Production and Health. *International Journal of Livestock Research*, 7(5), 1–7.

[33] Sammour, R. (2021). Cottonseed Oil: Production, Benefits, and Uses. In *Vegetable Oils: Types, Production, and Applications* (First Edition), 95–112. https://doi.org/10.1016/B978-0-12-820400-8.00005-4

[34] Sarkar, A. and Singh, B. (2021). Diversifying Agriculture: Exploring Non-Food Uses of Crops for Sustainable Development. *Current Science*, 120(5), 741–743.

[35] Sarkar, D. (Ed.). (2020). *Bioenergy and Biofuels from Agricultural Biomass*. CRC Press.

[36] Schoenlechner, R. and Berghofer, E. (2020). Wheat Bran as a Bioactive Component in Functional Foods. In *Wheat Quality for Improving Processing and Human Health* (First Edition), 239–258. https://doi.org/10.1016/B978-0-12-816360-3.00011-7

[37] Sharma, R., Kumar, V., Thakur, M. and Sharma, V. (2019). Medicinal and Aromatic Plants: A Boon for Sustainable Agriculture and Herbal Product Development. *Journal of Pharmacognosy and Phytochemistry*, 8(6), 3269–3275.

[38] Siddiq, M. and Sikandar, S. (2018). Groundnut Oil: From the Kernel to the Table. In *Nuts and Seeds in Health and Disease Prevention* (Second Edition), 543–550. https://doi.org/10.1016/B978-0-12-812130-4.00057-6

[39] Singh, Y. and Sidhu, B.S. (2019). Rice Straw Management: Current Status and Future Research Needs. In *Sustainable Agriculture Reviews* (First Edition), 243–261. https://doi.org/10.1007/978-3-030-16560-9_9

[40] Sivakumar, V. and Pandey, A. (2014). Biopolymers and Biotech Admixtures for Eco-Efficient Construction Materials. *Springer*.

[41] Sohaib, M., Anjum, F.M., Anwar, F. and Nawaz, H. (2019). Wheat Germ Oil: Extraction, Composition, and Applications. In *Oils and Fats in the Food Industry* (First Edition), 309–324. https://doi.org/10.1016/B978-0-12-804551-8.00019-8

[42] Sujatha, M. and Reddy, T.P. (2015). Castor. In *Genetic and Genomic Resources for Grain Crops Improvement* (pp. 299–324). Elsevier.

[43] Sujatha, M., Bhattacharya, A., Prasad, M.S. and Reddy, P.S. (2019). Current Status and Future Prospects of Castor Improvement: An Agronomic Perspective. *Indian Journal of Genetics and Plant Breeding*, 79(1), 1–14.

[44] Zeronian, S.H. and Collins, P. (2001). The Science and Technology of Cellulose Esters: Cotton Linters as a Source of Cellulose. In *Comprehensive Polymer Science and Supplements* (First Edition), 291–319. https://doi.org/10.1016/B978-008042699-3/00121-7.

Enhancing Soil Carbon Sequestration through Natural Farming Practices: A Sustainable Solution for Climate Change Mitigation and Food Security

C.K. Timbadia[1]

Introduction

The growing global population's high demand for food is compelling the agricultural sector to adopt advanced technologies, replacing traditional practices. Consequently, the sustainability of crop production systems, which relies on soil quality was being impacted by the farming methods to be employed. For instance, Intensive crop cultivation using imbalanced fertilizer, high nutrient mining through monoculture, excessive tillage and inversion coupled with the removal of crop residues by burning, hastens the decomposition of soil organic matter. This process can result in significant soil carbon loss, ranging from 20% to 67%, Yang *et al.* (2009). Such practices contribute to soil degradation leading to diminished physical, chemical, and biological properties Lal *et al.* (2014). Significant loss of soil organic carbon (SOC) ranging from 42% and 59% were reported when there is a transition in land-use patterns, such as from forest to crop and from pasture to crop respectively Guo *et al.* (2019). Overall, agricultural activities are responsible for emitting substantial quantities of greenhouse gases (GHGs) into the atmosphere, including carbon dioxide (CO_2), methane (CH_4), and nitrous oxide (N_2O) Smith *et al.* (2007). Natural Farming is increasingly advocated as an alternative approach to combat soil degradation caused by conventional agricultural practices that deplete soil fertility, with the aim of achieving higher crop productivity as a short-term benefit Kassam *et al.* (2009). Natural Farming practices are recognized to improve SOC contents in Natural Farming involves fundamental principles such as minimal soil disturbance, maintaining permanent soil cover or using cover crops, practicing mixed cropping, mulching with crop residues, utilizing on-farm

[1] Vice Chancellor, Gujarat Natural Farming Science University, Halol, Gujarat, India.

bio-inputs such as inoculum cultures of beneficial microbes and fermented botanicals for pest control. Therefore, addressing the dual challenges of food insecurity and climate change can be achieved by restoring soil carbon through the adoption of Natural Farming practices.

Causes of Carbon Loss from Soil

The loss of carbon (C) from the SOC pool occurs primarily in the form of carbon dioxide (CO_2) and methane (CH_4). Environmental factors, such as an increase in soil temperature, mainly stimulate the rate of mineralization of the SOC pool. Additionally, calciferous materials are subjected to certain climatic factors, leading to the dissolution of carbonates and bicarbonates, which releases CO_2 into the atmosphere. Burning of crop residues, monocropping, summer fallow, intensive cropping, excessive tillage and water deficiency are the key factor responsible for the loss of SOC.

Sequestration of Soil Carbon

The government advocates for the sequestration of soil organic carbon (SOC) to enhance food security and combat climate change at various forum. Soil is the best source of carbon sequestration absorbing CO_2 from Atmosphere it manage properly. According to this initiative, anthropogenic GHG emissions should be offset by an annual increase of global soil carbon stocks in the top 40 cm of soils by 0.4%. Additionally, conventional agricultural activities and land-use changes may contribute to GHG emissions, including approximately 25% of CO_2, 50% of CH_4, and 70% of N_2O emissions, which could potentially be balanced by SOC sequestration Hutchinson *et al.* (2007). Soil carbon sequestration involves transferring atmospheric CO_2 into soil organic matter (SOM), where carbon held in recalcitrant forms is less prone to losses through decomposition. Thus, SOC sequestration aims to retain captured atmospheric CO_2 in the slow SOC pool. However, it's recognized that the stable pool has limited potential for carbon sequestration due to its resistance to change from management practices Kane *et al.* (2015). In the short term, focusing on managing easily decomposable SOM is crucial. This can be achieved by increasing cropping intensity, which significantly impacts microorganisms and humic complex production, ultimately facilitating carbon sequestration. Carbon sequestration in soil can be achieved through below mentioned Natural Farming Practice:

1. Promoting soil microbial diversity and abundance by application of *Jeevamrit* and *Ghanjeevamrit.*

2. Residue mulching create favourable condition and supply food for microbes and earthworms.
3. Sustaining continuous living plant cover on soils throughout the year for root exudates.
4. Minimizing soil disturbance to improve the physical protection of soil carbon within aggregates.

Carbon Stabilization by Physical Protection in Aggregates

In most soils, the formation of young and unstable macroaggregates is aided by biological processes such as root growth, fungal and bacterial activity and the actions of soil fauna. These processes involve the mixing of fresh organic matter with soil particles and exudates. As partially decomposed organic matter becomes encapsulated within macroaggregates along with clay minerals and microbial products, microaggregates are formed. This leads to the long-term stabilization of carbon within macroaggregates, as it is shielded from rapid mineralization. Tiny particles of carbon, such as partially decomposed plant residues, become trapped within the center of aggregates. These particles are physically shielded from microbial degradation because microbes are unable to penetrate the core of these stable aggregates, where oxygen and water levels are low. As a result, microbial metabolism is inhibited in this environment Six *et al.* (2000). Hassink, *et al.* (2007). Research indicates that the turnover time of carbon is longer in microaggregates (412 years) compared to carbon in macroaggregates (140 years) Jastrow *et al.* (2006). This difference is attributed to the greater physical protection of organic matter within different aggregate-size classes, which depends on the quantity and composition of clay in the soil.

Impact of Tillage on Soil Organic Carbon

Conventional tillage practices lead to the disruption of soil aggregates in surface layers and increases aeration, which reduces total carbon by microbial decomposition of soil organic carbon to CO_2 particularly in macroaggregates. Additionally, adoption of zero tillage reduces the number of micro pores (15–150 μm) in the soil, which is crucial for microbial activity thereby improving the physical conservation of soil organic carbon. Therefore, by minimizing soil disturbance through the adoption of zero tillage practices, it is anticipated that CO_2 emissions from the soil to the atmosphere can be reduced, contributing to the mitigation of global climate change and improve soil organic carbon status. Furthermore, in a study under zero tillage, an average sequestration of 570000 grams of organic carbon/ha/year was recorded up to a soil depth of 30 cm. West and Post (2022).

Impact of Residue Mulching on Soil Organic Carbon

Currently, the widespread practice of burning crop residues is employed for managing stubble loads, leading to persistent issues such as nutrient loss, air pollution, and deterioration of soil health. Moreover, this practice contributes to a decline in SOC, as evidenced by a field trial conducted over 19 years in South-Eastern Australia, which reported a substantial loss of 1 lakh 75-thousand-gram carbon/ha/year in the 0–10 cm soil layer (Poeplau *et al.*, 2015). Conversely, according to Singh *et al.* (2016), the application of residue mulch treatments resulted sequestration of notable SOC content from 0.45% to 0.55%. However, retaining crop residues has been shown to enhance SOC content, particularly in the initial two decades, with benefits diminishing over the long term (Kirkby *et al.*, 2016). The application of residue mulching in crops creates favourable conditions and supplies food for microbes and earthworms. Additionally, the activity of root exudates in the rhizosphere, along with on farm bio-inputs, enhances degradation of residue to form humus organic substance resistant to degradation. Humus can make plant nutrients available and can hold 80-90% of water of its weight.

Agroforestry and Soil Carbon Sequestration

In addition to providing agricultural crops, fodder, and firewood/timber, agroforestry systems support numerous environmental benefits and ecosystem services. These include erosion control, improved water availability, increased species diversity, enhanced aesthetics of agricultural landscapes, and improved soil fertility through SOC sequestration. Moreover, these systems contribute to carbon fixation in tree biomass and the deposition of carbon-containing materials in both top soils and subsoils. They also result in decreased decomposition of resistant litter, reduced soil disturbance, and enhanced physical protection of organic matter through aggregates Hassink *et al.* (2013).

Conclusion

In conclusion, adopting sustainable agricultural practices such as Natural Farming, residue mulching, zero tillage, and agroforestry can mitigate soil degradation and enhance soil organic carbon levels. These practices promote soil health, reduce CO_2 emissions, and improve carbon sequestration, contributing to food security and climate change mitigation. Prioritizing soil conservation and carbon sequestration is crucial for building resilient agricultural systems that benefit both people and the environment.

References

[1] Guo, L.B. and Gifford, R.M. (2019). Soil carbon stocks and land-use change: A metaanalysis. *Global Change Biology*, 8(4):345–360.

[2] Hassinkm J. (2007). The capacity of soils to preserve organic C and N by their association with clay and silt particles. *Plant and Soil.* 191:77–87

[3] Hutchinson J.J., Campbell C.A. and Desjardins R.L. (2007) Some perspectives on carbon sequestration in agriculture. *Agricultural and Forest Meteorology*, 142:288–302

[4] Jastrow, J., Miller, R. and Boutton, T. (2006). Carbon dynamics of aggregate associated organic matter estimated by carbon-13 natural abundance. *Soil Science Society of America Journal*, 60:801–807.

[5] Kane, D. and Solutions, L. (2015). Carbon Sequestration Potential on Agricultural Lands: A Review of Current Science and Available Practices. *National Sustainable Agriculture Coalition Breakthrough Strategies and Solutions LLC*; 2015

[6] Kassam, A., Friedrich, T., Shaxson, F. and Pretty, J. (2009). The spread of conservation agri-culture: Justification, sustainability and uptake. *International Journal of Agricultural Sustainability*, 7:292–320.

[7] Kirkby, C.A., Richardson, A.E., Wade, L.J. Conyers, M. and Kirkegaard, J.A. (2016). Inorganic nutrients increase humification efficiency and C-sequestration in an annually cropped soil. PLoS One. 11: e0153698.

[8] Lal, R. (2014). Societal value of soil carbon. *Journal of Soil and Water Conservation*, 69(6):186A–192A.

[9] Poeplau, C., Katterer, T., Bolinder, M.A., Borjesson, G., Berti, A. and Lugato, E. (2015). Low stabilization of aboveground crop residue carbon in sandy soils of Swedish long-term experiments. Geoderma, 237:246–255.

[10] Singh, V.K., Dwivedi, B.S., Singh, S.K., Majumdar, K., Jat, M.L., Mishra, R.P. and Rani, M. (2016). Soil physical properties, yield trends and economics after five years of conservation agriculture-based rice-maize system in North-Western India. *Soil and Tillage Research*, 155: 133–148.

[11] Six, J., Elliott, E.T. and Paustian, K. (2000). Soil macroaggregate turnover and microaggregate formation: A mechanism for C sequestration under no-tillage agriculture. *Soil Biology and Biochemistry*, 32(14): 2099–2103.

[12] Smith, P., Martino, D., Cai, Z., Gwary, D., Janzen, H., Kumar, P., *et al.* (2007). Climate Change Mitigation. Contribution of Working Group III

to the Fourth Assessment Report of the Intergovernmental Panel on Climate Change. Cambridge, United Kingdom and New York, NY, USA: Cambridge University Press.

[13] Stavi I. and Lal R. (2013). Agroforestry and biochar to offset climate change: A review. *Agronomy for Sustainable Development*, 33:81–96.

[14] West, T.O. and Post, W.M. (2002). Soil organic carbon sequestration rates by tillage and crop rotation: A global data analysis. *Soil Science Society of America Journal*, 66:1930–1946.

[15] Yang, Y., Tilman, D., Furey, G. and Lehman, C. (2019). Soil carbon sequestration acceleration by the restoration of grassland biodiversity. *Natural Communications*, 10(1):1–7.

Agroforestry in the Context of Climate Change for Future Agriculture

A. Arunachalam[1], S. Suresh Ramanan[2] and A.K. Handa[3]

Introduction

Agroforestry has indeed been recognized as a distinct field of research since the 1970s. It emerged as a response to the recognition that traditional agriculture, which often involves monoculture cropping, can have negative environmental impacts such as soil erosion, loss of biodiversity, and reduced resilience to climate change. Agroforestry seeks to integrate trees and shrubs into agricultural systems to enhance their productivity, sustainability, and resilience. While agroforestry has been practiced for centuries in various forms around the world, it gained formal recognition and began to be studied systematically by scientists in the 1970s. Since then, research in agroforestry has expanded significantly, exploring various aspects such as silvopastoral, alley cropping, windbreaks, and other practices that integrate trees with agriculture.

Over the past few decades, interest in agroforestry has grown as scientists, policymakers, and farmers recognize its potential to address multiple challenges, including food security, climate change mitigation, biodiversity conservation, and rural livelihood improvement. Agroforestry research continues to evolve, with ongoing efforts to refine practices, assess their impacts, and promote their adoption on a larger scale.

The recognition of agroforestry's ecological benefits by organizations like the Intergovernmental Panel on Climate Change (IPCC) and other international bodies has indeed played a significant role in increasing its acceptance as a land-use practice. The emphasis on the ecological implications of agroforestry, including its potential to sequester carbon, enhance biodiversity, conserve soil and water, and provide habitat for

[1] Director, ICAR-Central Agroforestry Research Institute, Jhansi 284003, Uttar Pradesh, India.

[2] Scientist, M.Sc. (Agroforestry) ICAR-Central Agroforestry Research Institute, Jhansi, India.

[3] Principal Scientist (Forestry/Agroforestry), ICAR-Central Agroforestry Research Institute, Jhansi, India.

wildlife, has drawn attention from policymakers, researchers, and practitioners worldwide.

Furthermore, the acknowledgment of agroforestry's ability to provide multiple benefits across various sectors, including agriculture, livestock, and pastoralism, has contributed to its growing acceptance. Agroforestry systems can improve soil fertility, provide fodder for livestock, regulate water cycles, and offer shade and shelter, thereby enhancing the resilience of farming systems to environmental stresses and climate change impacts. The cross-sectoral benefits of agroforestry make it an attractive option for sustainable land management, particularly in regions vulnerable to climate change and land degradation. As a result, there has been continued global attention towards promoting agroforestry as a viable approach for addressing multiple challenges facing agriculture, the environment, and rural livelihoods.

Efforts to mainstream agroforestry into agricultural policies and practices have been increasing, with initiatives aimed at raising awareness, providing technical support, and incentivizing farmers to adopt agroforestry practices. This trend reflects a growing recognition of agroforestry's potential to contribute to sustainable development goals, including poverty alleviation, food security, and climate change mitigation and adaptation.

Institutionalization of Agroforestry Research in World

The systematic development of agroforestry research can be traced back to the initiatives of the Canada's International Development Research Centre (IDRC), Canada and FAO. While there was debate and deliberation on the relevance of agroforestry in temperate countries. In the early 1970s, agroforestry emerged as a distinct field of research, spurred by growing concerns about the environmental impacts of conventional agriculture. This led to the establishment of the International Council for Research in Agroforestry (ICRAF) in 1978, as a global research organization dedicated to agroforestry. Initially headquartered in Nairobi, Kenya, ICRAF aimed to promote sustainable land management practices and rural development through the integration of trees into agricultural landscapes. In 1982, ICRAF underwent a name change and became known as the World Agroforestry Centre (ICRAF), reflecting its expanded mandate and global reach. Under this new identity, ICRAF continued to advance agroforestry research, capacity building, and knowledge dissemination worldwide. In 2019, ICRAF merged with the Center for International Forestry Research (CIFOR) creating a unified research organization focused on addressing global challenges related to agriculture, forestry, and biodiversity conservation. This merger brought together complementary expertise and resources from both institutions, strengthening their collective capacity to support sustainable development and resilience in tropical landscapes. Today, the legacy of ICRAF lives on through the work of CIFOR and its partners in promoting agroforestry and sustainable land management practices around the world. In 1987, the Association for Temperate Agroforestry (AFTA) was established, dedicated to promoting research and education on agroforestry specifically in temperate regions. Since its inception, AFTA has played a pivotal role in advancing agroforestry knowledge and practices, facilitating collaboration among researchers, policymakers, and practitioners, and advocating for the integration of agroforestry into sustainable land management strategies tailored to temperate climates. Then, in 2009, the CGIAR Research Program on Forests, Trees, and Agroforestry (FTA) was inaugurated. This program marked a significant milestone in consolidating global research efforts in agroforestry, leading to notable advancements in agroforestry knowledge and the development of innovative practices. FTA's contributions have extended to addressing pressing challenges such as climate change mitigation and adaptation through agroforestry approaches. Furthermore, in 2014, the Global Landscapes Forum (GLF) launched the Agroforestry Knowledge Hub. This initiative has provided a vital platform for disseminating research findings, sharing best practices, and offering policy insights on agroforestry.

Institutionalisation of Agroforestry Research in India

The Indian Council of Agricultural Research (ICAR) organised a National Agroforestry Seminar at Imphal in 1979 which led to the formulation of All India Coordinated Research Project on Agroforestry in 1983 (Figure 1). Initially, this project was aimed at scientific enquiry and analysis of existing tree-based land-use systems and with subsequent findings from this project a dedicated research institute for agroforestry was created in 1988 – the National Research Centre for Agroforestry (currently known as the ICAR-Central Agroforestry Research Institute). In simple words, the institute is mandate with agroforestry research, extension and training. One of the prominent scientists in agroforestry, Dr. P.K.R. Nair says that the period between 1977 to 1987 as the decade of agroforestry research and concept development and institutionalization at global level. It will be apt to say that during same period agroforestry has been institutionalised here in India also.

The ICAR-Central Agroforestry Research Institute coordinates the All India Coordinated Research Project (AICRP) on Agroforestry from 1997 with the following objectives:

- Screening and genetic upgrading of selected plant species for their compatibility in different agroforestry systems.
- To optimize tree-intercrop combination for different regions.
- Performance enhancement of the pre-dominant agroforestry systems being already practiced by the farmers.
- To upgrade and refine the existing technologies for higher productivity and sustainability.

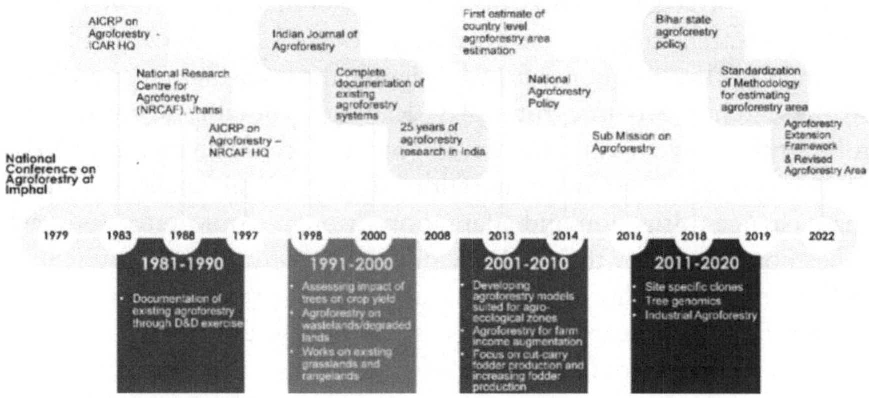

Figure 1: Agroforestry Research Roadmap

Mitigation Potential of Agroforestry to Climate change

Agroforestry is a viable alternative to prevent and mitigate climate change and has been recognized by IPCC as having high potential for sequestering C as part of climate change mitigation strategies. It can increase and stabilize agricultural yields and reduce soil erosion. The biomass produced through agroforestry can provide fuelwood, foods, basic construction materials, shade, medicines, etc., and thereby decreases pressure on natural forests. Furthermore, it may allow land to be taken out of fallow rotation in shifting cultivation systems; for example, one hectare of land sustainable managed with agroforestry could replace 5–10 ha of land under shifting rotation slash and burn. Due to this importance of agroforestry in the climate emergency era, the contribution of agroforestry is well recognized since its inception of systematic study of agroforestry in the 1970's.

Using trees as means of mitigating climate change can be achieved by maintaining the existing once on the farmlands or by increasing the plantation of short rotation and fast-growing trees in farmer's field. Agroforestry interventions, because of their ability to provide economic and environmental benefits, are considered to be the best measures in making farming communities adapt and become resilient to the impacts of climate change. Agroforestry practices have the greatest potential for conserving and sequestering carbon because of the close interaction between trees, crops, pasture, and soil. The important elements of agroforestry systems that can play a significant role in the adaptation to climate change include changes in the microclimate, protection through provision of permanent cover, opportunities for diversification of the agricultural systems, improving efficiency of use of soil, water and climatic resources, contribution to soil fertility improvement, reducing carbon emissions and increasing sequestration, and promoting gender equity. Two major aspect of mitigation strategies are carbon sequestration as well as reduced greenhouse gas emissions. In this context, the agroforestry systems sequester CO_2 from the atmosphere and contributes in the GHG emission reduction (Torres *et al.*, 2017). Palm *et al.* (2002) estimated that Nitrogen oxide (N_2O) emissions from tree-based agroforestry systems (9.8 μg N m^{-2} h^{-1} in Peach palm agroforestry) in Peruvian Amazon were about three times less than that of high input cropping systems (31.2 μg N $m^{-2}ha^{-1}$). Arid agroforestry can sequester carbon at a rate of 0.26 Mg C ha^{-1} yr^{-1} at a tree density of 9.71 ha^{-1} in the state of Rajasthan (India) and this contributes in the reduction of GHG emission of 1.42 million tonnes annually. N_2O emission in the agroforestry system was as estimated as low as 3%, CO_2 emissions were 70% of the high input cropping systems, and CH_4 uptake was nearly two-times that of the low input cropping system in of Subtropical China (Wang *et al.*, 2015). Gross *et al.* (2021) measured that N_2O emissions were 89% lower under agroforestry system compared to the cropland (0.02 and 0.18 g Nm^{-2} $year^{-1}$, respectively) in central Alberta, Canada.

Similarly, with regard to carbon sequestration through agroforestry, a study reported thebaseline standing biomass in the tree componentsvaried from 2.45 to 2.88 Mg dry matter (DM) ha^{-1} and the totalbiomass (tree-crop) from 11.14 to 25.97 MgDM ha^{-1} in the three districts. The soil organic carbonin the baseline ranged from 8.13 to 9.12 Mg C ha^{-1} and is expected to increase from 8.63 to 24.51 Mg C ha^{-1}. The CSP of existing agroforestry systems (for 30 years simulation) has been estimated to the tune of0.111, 0.126 and 0.551 Mg C ha^{-1} $year^{-1}$ for Sultanpur, Dinajpur and Ludhiana districts, respectively (Ajit *et al.* 2013). Over the period of time, the various studies have reported the carbon sequestration through agroforestry. However, the sequestration potential estimated through agroforestry for Indo-Ganetic plains alone was about 36.25 t C ha^{-1}.

Adaptation of Agroforestry to Climate Change

To address the need to adaptability of agroforestry systems to climate change, the screening and genetic upgrading of selected tree species for their compatibility in different agroforestry systems have been carried out. Owing to the systematic efforts, about 184 promising tree species germplasms were collected and are being evaluated for its superiority. In this regard, registration of the elite germplasm has been done like shisham by NRCAF (Bundel-1 and Bundel-2) and GBPUAT, Pantnagar (PS 52), poplar clones (L-48/89, L-47/88) by PAU, Ludhiana, Pant Poplar by GBPUAT, Pantnagar, teak clone (PDKV/AF-1) by College of Agriculture, Nagpur and eucalyptus (SRY-16) by MPKV, Rahuri. Similarly, in neem, elite germplasm with high yield and high, stable azadirachtin content have been identified and are further explored for genetic gains. The AICRP on Agroforestry centres have been also exploring new species to be incorporated into agroforestry systems. With industrial agroforestry and contract farming gaining popularity, fast growing species like *Melia dubia, Anthocephalus cadamba and Melia azedarach* were focused in recent years and promising clones like Malabar Neem (*Meliadubia*) - MTP 1, MTP 2 & MTP 3; Kadam (*Anthocephaluscadamba*) - MTP 2 by TNAU centre; and *Melia azedarach* - Punjab Dek 1 & Punjab Dek 2 by PAU centre was also released. Agroforestry research does not focus on timber yielding trees alone, NTFPs trees were also screened for superior genetic gains and clones/varieties like Undi (*Calophylluminophyllum*) clone KKVCI-03 by BSKKV centre; Imli (*Tamarindusindica*) varieties viz., DTS-1 and DTS-2 by UAS Dharwad centre; and GKVK-17 Tamarind variety for commercial cultivation to Eastern Dry of Karnataka was also released recently.

As the objective is also to screen plant species for their compatibility in different agroforestry, the AICRP on Agroforestry centres have also screened crop varieties suitable for specific agroforestry systems (models). For instance, Wheat varieties WH 1105, PBW 677, PBW 725, PBW

502, DBW 17, PBW 550 and PBW 621 are suitable for Popular based agroforestry system in Punjab region. The findings from RPCAU centre state that Krishna-258, a Til (*Sesamumindicum*) variety is superior and suitable for intercropping up to 5 years in the Shisham (*Dalbergiasissoo*) based agroforestry system in Bihar.

The continuous effort to the AICRP on Agroforestry in the past 40 years has translated to develop agroforestry system (models) specific different agro-ecologcial reasons of the country. For instance, in Deccan Plateau having 600–1000 mm rainfall can adopt a) Three-tier Agroforestry System for Paddy Growing Area with Teak and Mango as Tree component and Paddy (Kharif); Gram, Black gram, Linseed, Lathyrus (Kharif) as crop component; b) Sapota-Teak based Agroforestry System for Hill Zone of Karnataka with Teak and Sapota as Tree component and Paddy (Kharif); South African Maize, Sun hemp (Kharif) as crop component; and c) Tamarind based Silvi-horticultural System with *Tamarindusindica*, Eucalyptus and Casuarina as Tree component and Natural grass (DTS-1, DTS-2 and SMG-13 as crop component for pasture/fodder. Similarly for specific agroforestry systems for all the 20 agro-ecological zones along with their economic analysis have been developed for the country.

India has always been a pioneer in estimating the area under agroforestry. Earlier attempts at the country level revealed estimates varying from 17.45 to 23.25 million ha and many regional estimates are also reported. There are papers predicting the potential area suitable for agroforestry in India. Despite the predictions, there are no actual estimates to date. ICAR-Central Agroforestry Research Institute (CAFRI), a dedicated research institute for agroforestry in the Asia-Pacific region took up the mapping of agroforestry areas using geospatial technologies. The preliminary work on 13 out of 15 agro-climatic zones reported an area of 23.25 m ha (Rizvi *et al.*, 2019). In 2022, we carried out the complete analysis for all 15 agro-climatic zones of India.

The overall area under agroforestry area for all 15 agro-climatic zones of India works out to be 28.427 million ha, which is about 8.65 per cent of the total geographical area of the country (328.762 m. ha.) (Table 1). Out of 15 agro-climatic zones, seven zones (1, 3, 5, 7, 11, 12 & 13) have more than 10 per cent area under agroforestry. Agro-climatic zones 1, 5, 7, 10, 11 and 13 have more than 2 million ha of the area under agroforestry.

Table 1: The Extent of Agroforestry Area in 15 Agro-Climatic Zones of India

ACZ No.	*Agro-climatic Zones*	*Geographical Area (M ha)*	*Agroforestry Area (M ha)*	*Agroforestry Area (%)*
1	Western Himalayan Region	32.968	4.096	12.42
2	Eastern Himalayan Region	28.422	1.088	3.83
3	Lower Gangetic Plains Region	6.238	0.802	12.86
4	Middle Gangetic Plains Region	16.526	1.304	7.89
5	Upper Gangetic Plains Region	14.367	2.234	15.55
6	Trans Gangetic Plains Region	11.750	1.143	9.73
7	Eastern Plateau & Hill Region	40.525	4.292	10.59
8	Central Plateau & Hill Region	37.435	1.924	5.14
9	Western Plateau & Hill Region	32.539	1.556	4.78
10	Southern Plateau & Hill Region	39.294	2.976	7.57
11	East Coast Plains & Hill Region	19.948	2.36	11.83
12	West Coast Plains & Hill Region	11.69	1.632	13.96
13	Gujarat Plains & Hill Region	18.673	2.57	13.76
14	Western Dry Region	17.587	0.431	2.45
15	The Island Region	0.785	0.019	2.42
Total/Percentage		328.747	28.427	8.65

The Northern Himalayan Region and the Eastern Plateau & Hilly Region recorded more than 4 million ha of the area under agroforestry. Across the agro-climatic zones proportionately, the Upper Gangetic Plain Region had a greater area (15.55%) under agroforestry and lowest in the Western Dry Region (2.45%) and the Island region (2.48%). As per the Global Forest Resources Assessment – 2020 (FAO 2020), Asia has the largest area under agroforestry which is about 31.2 m ha. In comparison with the result of this work, it can be presumed that more than 75% agroforestry area is in India. However, it is not a fact. Globally, only 71 countries are reporting areas under agroforestry to FAO for the biannual global forest resources assessment but there is no actual estimation reported.

With this, India becomes the first country to have mapped the country-wide agroforestry area mapping systematically. As per ISFR 2019 (FSI

2021), the extent of the Trees Outside Forests area is 29.38 mha *i.e.*, 9.5 mha of tree cover and 19.88 m ha of forests outside the reserved forest area. Thus, the agroforestry area estimate is inclusive of the 9.8 mha of tree cover (rural and urban). Moreover, the agroforestry area reported here indicates both tree and crop canopy areas. Hence, it is advised not to equate the agroforestry area as such to the trees outside the forests area.

Relevance of Agroforestry for the Sustainable Developmental Goals (SDGs)

The central focal point of all SDGs is the "land-use" (van Noordwijk *et al.*, 2018). Agroforestry as a land-use can be proven to follow all the principles of the sustainable land management. As per the World Overview of Conservation Approaches and Technologies (WOCAT), the key principles are buildup of soil organic matter and related biological activity, integrated plant nutrition management, better crop management, better rainwater management, improvement of soil rooting depth and permeability and reclamation of the degraded land (Sanz *et al.*, 2017). Agroforestry adheres to these principles perfectly. However, agroforestry is not that competent enough in sustainable land management principles compared to forestry. And also, on comparison between agriculture, forestry and agroforestry in contribution for achieving SDGs, forests are better performing in SDG 13 (Climate change) and 15 (Life on Land) and agroforestry is the second-best option (Figure 2). However, the

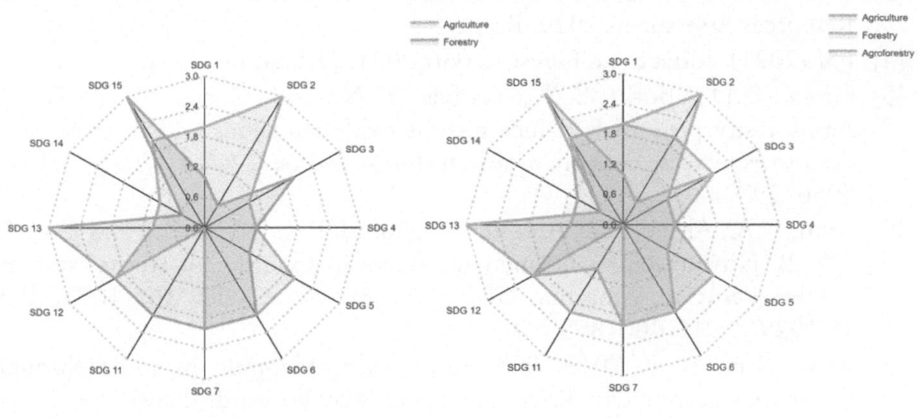

Figure 2: Comparing Relevance Level and Contribution of Agriculture, Forestry and Agroforestry Sector in Achieving SDGs

contribution of forestry in achieving SDG 2 (Zero Hunger) is minimal and agriculture plays a major role in achieving SDG 2 but its contribution in other SDG is lower compared to forestry. In contrast to both forestry and agriculture sectors, agroforestry plays a substantial role in achieving 12 out of 17 SDGs. Moreover, in countries like India where land is scarce resource, more land cannot be diverted for forestry sector. Agroforestry as a multifunctional landuse pattern forms a mosaic landscape supporting rich biodiversity inside the matrix of agricultural lands.

Acknowledgement

The authors thank the Indian Council of Agricultural Research, New Delhi and the DST-National Mission on Sustaining Himalayan Ecosystems (NMSHE), Govt. of India for all support.

References

[1] Gupta, Ajit; Dhyani, S.; Ramnewaj, Handa, A.K.; Prasad, Rajendra; Alam, Badre; Rizvi, Raza; Gupta, Garima; Pandey, K. and Jain, Dr. Amit & Uma (2013). Modeling analysis of potential carbon sequestration under existing agroforestry systems in three districts of Indo-gangetic plains in India. *Agroforestry Systems,* 87. 10.1007/s10457-013-9625-x.

[2] Chavan, S.B., Newaj, R., Rizvi, R.H., Ajit, Prasad, R., Alam, B. and Tripathi, D. (2021). Reduction of global warming potential vis-à-vis greenhouse gases through traditional agroforestry systems in Rajasthan, India. *Environment, Development and Sustainability*, 23, 4573–4593.

[3] FAO (2020). Global Forest Resources Assessment 2020 Global Forest Resources Assessment 2020. Rome.

[4] FSI (2021). India State Forest Report 2021. Dehradun, India.

[5] Gross, C.D., Bork, E.W., Carlyle, C.N. and Chang, S.X. (2022). Agroforestry perennials reduce nitrous oxide emissions and their live and dead trees increase ecosystem carbon storage. *Global Change Biology*, 28(20), 5956–5972.

[6] Palm, C.A., Alegre, J.C., Arevalo, L., Mutuo, P.K., Mosier, A. and Coe, R. (2002). Nitrous oxide and methane fluxes in six different lan use systems in the Peruvian Amazon. *Global Biogeochem.* Cycles 16: 1073, doi: 10.1029/2001GB001855.

[7] Rizvi, R.H. *et al.* 2019. 003 Agroforestry Mapping in India through Geospatial Technology: Present Status & Way Forward. Jhansi.

[8] Sanz, M.J. *et al.* (2017). "Sustainable Land Management Contribution to Successful Land-Based Climate Change Adaptation and Mitigation. A Report of the Science-Policy Interface." In United Nations Convention to Combat Desertification United Nations.

[9] Shrestha, B.M., Chang, S.X., Bork, E.W. and Carlyle, C.N. (2018). Enrichment planting and soil amendments enhance carbon sequestration and reduce greenhouse gas emissions in agroforestry systems: a review. *Forests*, 9(6), 369.

[10] Torres, C.M.M.E., Jacovine, L.A.G., Nolasco de Olivera Neto, S., Fraisse, C.W., Soares, C.P.B., de Castro Neto, F., ... and Lemes, P.G. (2017). Greenhouse gas emissions and carbon sequestration by agroforestry systems in southeastern Brazil. *Scientific Reports*, 7(1), 16738.

[11] van Noordwijk, Meine *et al.* (2018). "SDG Synergy between Agriculture and Forestry in the Food, Energy, Water and Income Nexus: Reinventing Agroforestry?" *Current Opinion in Environmental Sustainability,* 34: 33–42.

[12] Wang, G.; Welham, C.; Feng, C.; Chen, L. and Cao, F. (2015). Enhanced Soil Carbon Storage under Agroforestry and Afforestation in Subtropical China. *Forests* 6, 2307–2323.

Prospects of Digital Agriculture in Gujarat

A.R. Pathak[1], P.A. Pandya[2] and D.K. Varu[3]

Introduction

Agriculture is the backbone of Gujarat's economy, contributing significantly to its GDP and providing livelihoods to millions of people. Horticulture, in particular, has gained prominence due to its high-value crops and export potential. With the advent of digital technologies, the agricultural landscape in Gujarat is undergoing a profound transformation. Digital agriculture encompasses the use of various technologies such as precision farming, IoT (Internet of Things), remote sensing and data analytics to optimize crop production, minimize resource wastage and enhance farm efficiency. With the world population growing and resources becoming scarcer, the need for sustainable and efficient agricultural practices is paramount. Digital technology offers a suite of tools and solutions to address these challenges, enabling farmers to optimize resource use, minimize environmental impact, and enhance crop quality and yield. This paper examines the application of digital technology across various aspects of agriculture, illustrating its transformative potential and key considerations for adoption.

Digital Agriculture: Indian Context

In India, the pressing issue of low farmers' incomes necessitates efforts to enhance their livelihoods. A pivotal strategy for achieving this goal involves augmenting the efficiencies of agricultural production processes and the entire value chain. Globally, Digital Agriculture has emerged as a promising avenue to bolster efficiencies and boost incomes in agriculture. The concept of Digital Agriculture within the Indian context, while examining the challenges and potential opportunities is discussed.

---

[1] Former Vice Chancellor, Junagadh Agricultural University, Junagadh & Navsari Agricultural University, Navsari, Gujarat-362001, India.

[2] Assistant Professor, Junagadh Agricultural University, Junagadh, India.

[3] Professor, Junagadh Agricultural University, Junagadh, India.

India's recent focus on bolstering farmer incomes has gained significant traction, particularly following the Prime Minister's call to double farmers' income, subsequently echoed in official government policies. Despite India's commendable agricultural production achievements, ranking first in milk, jute, and pulses and second in wheat, rice, and other key crops, the primary concern remains the inadequate incomes of farmers, largely due to fragmented landholdings.

Digital Technologies offer a compelling solution to enhance agricultural productivity and profitability. The integration of technologies such as Artificial Intelligence (AI), Internet of Things (IoT), and image recognition holds immense potential across various facets of agriculture, from gene sequencing in seed production to precision management through data-driven insights.

The definition of Digital Agriculture varies among different stakeholders. Precision Agriculture, entails the strategic utilization of temporal, spatial, and individual data to optimize resource utilization, productivity and sustainability in agricultural production. Similarly, Smart Farming, also known as Farming 4.0, leverages information and data technologies to optimize complex farming systems, enabling tailored seed planting and informed decision-making based on concrete data.

For Digital Agriculture to thrive in India, several key factors are imperative. These include the affordability of technology, user-friendly portable hardware, flexible payment models such as pay-per-use renting, robust policy support, and leveraging the collective power of farmer cooperatives. Moreover, initiatives such as India's National Strategy on AI underscore the significance of integrating digital solutions in agriculture to unlock economic and social benefits.

Precision Agriculture, Plant Monitoring and Management

It involves the use of advanced technologies, such as GPS, GIS, drones, and sensors, to optimize farm management practices. Precision agriculture enables farmers to precisely monitor and manage factors such as soil moisture, nutrient levels and pest infestations. By collecting real-time data and using predictive analytics, farmers can make informed decisions about planting, irrigation, fertilization, and pest control, leading to improved yields and resource efficiency (Huang *et al.*, 2020).Digital technology enables real-time monitoring and management of crops throughout the growing season. Remote sensing technologies, such as satellites and drones, capture multispectral imagery to assess crop health, detect diseases, and monitor growth patterns. Machine learning algorithms analyze this data to identify stress factors and recommend interventions, such as adjusting fertilizer applications or implementing targeted pest control measures. Additionally, mobile apps and web-based platforms provide farmers with access to crop management tools, pest and disease databases, and decision support systems, facilitating proactive management practices.

Digital Irrigation for Crops

Efficient water management is crucial for successful agricultural production, and digital irrigation technologies offer innovative solutions to optimize water use and improve crop yields. Digital irrigation systems consist of several components, including sensors, actuators, controllers, and communication networks. Soil moisture sensors measure soil moisture levels at different depths, providing real-time data on plant water status and irrigation needs. Weather stations collect meteorological data, such as temperature, humidity, and rainfall, to inform irrigation scheduling and adjustments. Controllers use algorithms to analyze sensor data and automate irrigation operations, ensuring precise and efficient water application. Communication networks enable remote monitoring and control of irrigation systems, allowing farmers to make timely adjustments and optimize water use.

Water scarcity is a significant challenge in agriculture, particularly in arid and semi-arid regions. Smart irrigation systems leverage digital technology to optimize water use and minimize wastage. These systems incorporate sensors, weather data, and algorithms to automatically adjust irrigation schedules based on soil moisture levels, weather forecasts, and plant water requirements. By delivering the right amount of water at the right time, smart irrigation systems can improve crop health, reduce water consumption, and prevent soil erosion (Zhang *et al.*, 2019).

Digital Technologies in Soil Health Management

Soil health management is a critical component of successful production, influencing crop growth, yield, and quality. Digitalization, encompassing technologies such as precision agriculture, Internet of Things (IoT), and data analytics, is revolutionizing soil health management practices.

Digital technologies offer a range of tools and solutions for soil health management in agriculture. Precision agriculture techniques, including soil mapping, variable rate application, and site-specific management, enable farmers to optimize inputs and improve soil fertility. IoT devices such as soil sensors and weather stations provide real-time data on soil moisture, temperature, and nutrient levels, facilitating timely decision-making and resource allocation. Data analytics platforms analyze large datasets to identify trends, patterns, and correlations, informing soil management strategies and predictive modelling. By optimizing soil conditions and nutrient management practices, digitalization enhances crop productivity and quality while minimizing environmental impact. Improved monitoring and analysis of soil health parameters enable early detection of nutrient deficiencies, soil compaction, and erosion, allowing for proactive interventions to mitigate risks and optimize yields. Additionally, digitalization fosters data-driven decision-making and precision management, leading to more efficient resource use and reduced input costs.

Digital Mechanization in Agriculture: Revolutionizing Planting, Pruning, and Harvesting

Planting is a critical phase in agricultural production and digital mechanization offers several advancements to streamline this process. Precision planting technologies, including GPS-guided planters and automated seed dispensers, enable precise spacing, depth, and placement of seeds or seedlings. Robotics and automation systems further enhance efficiency by reducing human error and labour requirements. Real-time monitoring and feedback mechanisms ensure optimal planting conditions, resulting in improved germination rates and crop establishment.

In horticulture pruning is essential for shaping plant growth, improving fruit quality, and maximizing yields in horticultural cropss. Robotic pruners equipped with sensors and vision systems can accurately identify and remove unwanted branches while preserving healthy growth. Machine learning algorithms analyze plant morphology and growth patterns to customize pruning strategies for different plant varieties and growth stages. Automated pruning systems increase efficiency, reduce labor costs, and ensure consistency in pruning practices.

Harvesting is a labour-intensive and time-critical operation and digital mechanization offers transformative solutions to streamline this process. Robotics and automation technologies, such as robotic harvesters and automated picking systems, enable efficient and precise harvesting of fruits, vegetables, and other horticultural crops. Computer vision and machine learning algorithms identify ripe produce, assess quality parameters, and facilitate selective harvesting. Real-time data analytics optimize harvest schedules, logistics, and post-harvest handling, minimizing losses and maximizing marketable yields.

Digital Agro-Processing and Value Addition of Crops

Agro-processing and value addition play a crucial role in enhancing the market value and competitiveness ofl crops. With the advent of digital technologies, there is a growing interest in leveraging digital solutions to streamline agro-processing operations, improve product quality, and increase market access.

Digital technologies such as automation, robotics, artificial intelligence (AI), and Internet of Things (IoT) are revolutionizing agro-processing operations for crops. Automated sorting and grading systems use machine vision and AI algorithms to classify fruits and vegetables based on size, shape, colour, and quality parameters. IoT-enabled sensors monitor processing parameters such as temperature, humidity, and pH levels in real-time, ensuring optimal conditions for product quality and safety. Robotics and automation systems automate la packaging, improving efficiency and reducing processing time.

Digital platforms enable farmers and processors to access market information, consumer preferences, and trends, allowing them to develop and market value-added products that meet customer demands. Customization tools and algorithms tailor product attributes such as flavor, texture, and nutritional values. Digital branding and marketing strategies leverage social media, e-commerce platforms, and digital storytelling to create brand awareness, build consumer trust, and increase market share.

Digital Market Analysis and Supply Chain Management

The agriculture industry is experiencing a digital transformation, with advancements in technology reshaping market analysis and supply chain management practices. By analyzing the impact of digital solutions on supply chain efficiency, market transparency.

The digital agriculture market is witnessing rapid growth, driven by factors such as increasing adoption of precision agriculture, IoT devices, and data analytics. Market analysis tools leverage big data analytics and machine learning algorithms to provide real-time insights into market trends, consumer preferences, and competitive dynamics. Digital platforms enable stakeholders to access market information, pricing data, and demand forecasts, facilitating informed decision-making and strategic planning. Moreover, e-commerce platforms and online marketplaces offer new channels for producers to reach consumers directly, bypassing traditional intermediaries and reducing transaction costs.

Digital technologies such as blockchain, RFID, and GPS tracking systems enhance supply chain visibility, traceability, and transparency. Blockchain technology enables secure and immutable record-keeping of transactions and product provenance, reducing the risk of fraud and counterfeiting. RFID tags and GPS trackers provide real-time location tracking of products throughout the supply chain, enabling efficient inventory management, logistics optimization, and quality control.

Utilizing Drone

The utilization of drones in farming has brought about a significant transformation in traditional agricultural practices, offering multifaceted benefits to farmers. With their high-resolution cameras and sensors, drones enable precise monitoring and assessment of crops, soil conditions, and pest infestations, providing farmers with invaluable insights into crop health and growth trajectories (Singh *et al.*, 2020). This aerial perspective facilitates timely interventions and optimized resource allocation, thereby enhancing overall agricultural productivity (Singh & Pardeshi, 2019). In the context of India, where horticulture plays a vital role in agricultural diversification and income generation (Singh *et al.*, 2019), drones have proven particularly advantageous. They can be utilized for a range of tasks including crop scouting, mapping, and targeted spraying, aiding in the identification of areas requiring irrigation or fertilization and improving water and nutrient management practices (Rathore *et al.*, 2021). Moreover, drones contribute to the early detection of diseases and pest outbreaks, allowing for prompt mitigation measures to be implemented, consequently minimizing crop losses (Krishna *et al.*, 2018). Furthermore, by reducing the reliance on manual labor and precisely targeting inputs, drones help minimize environmental impact and operational costs in farming (Singh & Mishra, 2020). Thus, drones have emerged as indispensable tools in modern horticulture farming in India, playing a pivotal role in enhancing efficiency, productivity, and sustainability in agricultural practices.

Digital Agriculture and Gujarat

Use of ICT Tools and Social Media: The integration of Information and Communication Technology (ICT) and social media platforms is significantly impacting the operations of Farmer Producer Organizations (FPOs) engaged in agriculture in Gujarat. Given the evolving agricultural landscape, effective communication channels and access to pertinent information are essential for farmers to optimize productivity and expand market outreach. Across Gujarat, a considerable number of farmers have adopted ICT tools and social media platforms as fundamental elements of their agricultural activities. These technological advancements facilitate the dissemination of vital agricultural insights, market dynamics, and best practices, empowering both FPOs and individual farmers to make well-informed decisions regarding cultivation techniques, pest management strategies, market positioning, and more. By embracing ICT and social media, become more efficient and effective, but they also cultivate networking opportunities, foster knowledge exchange, and enhance access to resources, thereby contributing significantly to the sustainable growth of Gujarat's agricultural sector. Apart from the regular features of providing information regarding weather and government schemes, the 'Gujarat Sarkar Khedut Mitra' app allows farmers to interact with other farmers as well as government officials through the app. Patel *et al.* (2021) conducted a study examining the socio-economic status of farmers and their utilization of social media for sustainable agricultural development in Gujarat State. Their findings revealed that nearly half of the farmers (48.33%) exhibited a high level of engagement with ICT tools, with 40.84% reporting a medium level of usage and only 10.83% indicating low usage. WhatsApp emerged as the preferred platform, with 100% of farmers utilizing it, and 96.67% accessing it regularly. Additionally, YouTube (86.67%) and Facebook (80.00%) were commonly used by farmers for information gathering and entertainment. In a subsequent study by Patel *et al.* (2022) on predictive factors influencing farmers' knowledge of social media for sustainable agricultural development, it was found that 65% of farmers possessed a high level of understanding regarding social media usage. Factors such as education, occupation, income, innovativeness, scientific orientation, ICT tool usage, social media platform usage, and information-seeking behavior were positively and significantly correlated with farmers' knowledge of social media, while age and farming experience were negatively correlated. These findings underscore the importance of tailored ICT programs and farmer-centric social media platforms to capitalize on current trends in farmer behavior, offering valuable insights for policymakers engaged in farmer capacity building initiatives.

Smart Sensor Technology in Automated Irrigation Systems

The incorporation of smart sensor technology into automated irrigation systems has become indispensable in contemporary agricultural methods. These advanced tools offer a multitude of benefits, such as precise monitoring of soil moisture levels, real-time tracking of weather conditions and accurate assessment of crop water requirements. Gujarat state is located at the peripheral boundary of the southwest monsoon; hence, the distribution of rainfall is extremely uneven and irregular (Pandya and Gontia, 2023) Due to problems like water scarcity, erratic rainfall and contaminated groundwater in Gujarat, a notable proportion of farmers have embraced mobile applications integrated with smart sensors to optimize irrigation practices for their horticultural crops. Through the utilization of these innovative solutions, farmers can efficiently regulate water usage, minimize wastage, and amplify crop yields while preserving crucial resources. Moreover, the implementation of automated irrigation systems facilitated by smart sensors alleviates the manual burden of monitoring and decision-making, enabling farmers to allocate their attention to other farm management aspects. The widespread adoption of such technologies underscores their significance in advancing sustainable agricultural practices and bolstering productivity within Gujarat's agricultural sector. In their study, Patel *et al.* (2022) underscored the significance of smart irrigation systems tailored for shared Tubewell culture in the North Gujarat region. The proposed system is adept at delivering optimal water supply to crops by employing a sequencing algorithm based on various farm data parameters. Prior to initiating irrigation, the system proactively assesses rain predictions. In instances of high probability of rain, irrigation is deferred, thus preventing unnecessary water application. Conversely, in areas with lower rain probability, the system facilitates partial irrigation. Termed as the IAA Architecture (Irrigation Anytime from Anywhere Architecture), this proposed framework demonstrates a proactive approach towards efficient water management in agriculture. Several farmers in Gujarat are leveraging mobile-based applications such as Fasal Salah, which heralds a new era of agricultural support by providing unparalleled levels of personalization. Fasal Salah stands out as an exemplary mobile app designed to assist farmers by furnishing real-time, tailored crop advisories based on current and forecasted weather conditions. Offering precise weather forecasts encompassing temperature, humidity, wind speed and direction, and rainfall for the upcoming 10 days at both taluk and village levels, Fasal Salah emerges as a unique and farmer-friendly solution.

This innovative platform represents a significant advancement in the agricultural domain, equipping farmers with essential insights to make informed decisions, enhance crop yields, and effectively manage weather-related challenges. For instance, Mr. Kashyap Rudani from Village Devpar, Taluka Mandvi, Kutch, has embraced this application through his Farmers' Producer Organization (FPO) to optimize his pomegranate crop management. His experience reflects reduced water and pesticide usage, along with timely disease and pest management, showcasing tangible benefits.

Similarly, other farmers like Vasant Bhai L. Patel and his FPO members in Kutch utilize this Android-based application for cultivating crops such as pomegranate, dragon fruit, and guava. Additionally, Mr. Niteshbhai Mangnvala and his FPO members rely on Fasal Salah for cultivating crops like pomegranate, grapes, mangoes, and date palms. These testimonies underscore the transformative impact of Fasal Salah in empowering farmers across diverse agricultural landscapes, fostering sustainable practices, and maximizing yields.

Application of Drones in Horticulture

The Agricultural Technology Application Research Institute (ATARI) in Pune, Zone-VIII, oversees agricultural initiatives across Maharashtra, Gujarat, and Goa, housing a total of 82 Krishi Vigyan Kendras (KVKs), with 50, 30, and 2 KVKs respectively in the aforementioned states. Under the Agri Drone Project of the Sub-Mission on Agricultural Mechanization, the Ministry of Agriculture sanctioned 40 drones in the fiscal year 2022–23. Among these, 7 were allocated to State Agricultural Universities, 23 to ICAR Institutes, and 10 to Krishi Vigyan Kendras. To ensure competent operation, ATARI Pune, in collaboration with the Directorate General of Civil Aviation (DGCA), facilitated drone pilot training through 63 authorized Remote Pilot Training Organizations (RPTOs) nationwide. Among the implementing centers, Rashtriya Raksha University in Gandhinagar was designated as the RPTO for Gujarat. Additionally, centers such as AAU Anand, JAU Junagadh, SDAU Dantiwada, NAU

Navsari, ICAR-DGR Junagadh, and various KVKs across Gujarat underwent training programs. To date, 22 drone pilots have been trained in Gujarat as part of this initiative (Athare *et al.*, 2024). Furthermore, recognizing the potential of agricultural drones, the Gujarat state government has allocated Rs 35 crore for agri-drone projects aimed at covering 1.4 lakh acres of farmland, with a focus on assisting farmers in procuring drones (Anonymous, 2022). Department of Agriculture and farmers welfare, GoI will focus on promotion of drone in applications such as Spraying of Pesticides, Fertilizer Application, Seed Sowing, Soil Quality Monitoring, Survey of Soil Erosion.

Apart from this, following initiate are taken in Digital Agrticulture Development in Gujarat

- The state government, in collaboration with various stakeholders, has launched several initiatives to promote digital agriculturals practices. For instance, the "e-Krishi Yantra Anudan Yojana" provides subsidies to farmers for purchasing digital farming equipment and machinery.
- Similarly, the "Khedut Portal" offers online services related to agriculture, including information on crop cultivation, weather forecasts, and market prices. These initiatives have facilitated the widespread adoption of digital technologies among farmers in Gujarat.
- By leveraging IoT devices and sensor technology, farmers can monitor soil moisture levels, nutrient content, and pest infestations in real-time, allowing for timely interventions. This not only improves crop yields but also reduces input costs and environmental impact. Furthermore, digital technologies enable better market linkages, enabling farmers to access information on market demand and prices, thus maximizing their profits.
- Mobile applications have been developed in Gujarat by University as well as some private partners to provide farmers with weather forecasts, market prices, crop management tips, and access to agricultural experts for guidance and support.
- Encouraging the growth of agri-tech startups in Gujarat that focus on developing innovative digital solutions for agriculture, such as farm management platforms, remote sensing technologies, and farm-to-market platforms.
- These digital initiatives aim to empower farmers with data-driven insights, improve resource efficiency, reduce production costs, enhance market access, and ultimately contribute to the sustainable growth of the agriculture sector in Gujarat.

Challenges and Future Prospects

The adoption of digital agriculture faces several challenges in Gujarat. One of the primary challenges is the digital divide, with many smallholder farmers lacking access to technology and digital literacy. Limited internet connectivity and electricity supply in rural areas further exacerbate this divide. Additionally, high initial investment costs and the complexity of implementing digital solutions pose barriers to adoption for some farmers. Addressing these challenges requires concerted efforts from government, private sector and civil society organizations.Looking ahead, the future of digital agriculture in Gujarat appears promising. Advances in technology, such as AI (Artificial Intelligence) and machine learning, hold immense potential for further optimizing agricultural practices and decision-making. Moreover, initiatives aimed at bridging the digital divide, such as rural internet connectivity projects, will help ensure that all farmers can benefit from digitalization. By embracing digital agriculture, Gujarat can not only enhance its agricultural productivity and sustainability but also empower farmers and strengthen rural economy.

References

[1] Anonymous (2022). https://yourstory.com/2022/09/gujarat-government-allocates-rs-35-cr-agri-drone-scheme.

[2] Athare, T., T.R., Roy, S.K. and Chaware, G. (2024). Application of drone technology for sustainable agriculture in western India. *Indian Farming,* 74(3), 52–55.

[3] Brodie, J. and Dalglish, C. (2018). The blockchain revolution: An analysis of regulation and technology related to distributed ledger technologies. *Computer Law & Security Review,* 34(5), 1210–1222.

[4] Government of Gujarat. (n.d.). Agriculture & Co-operation Department. Retrieved from https://agri.gujarat.gov.in

[5] Huang, Y., Cui, Y., Cheng, T. and Zhu, M. (2020). A Review of the Applications of Precision Agriculture in Horticultural Crops. *Frontiers in Plant Science,* 11, 596097.

[6] Iqbal, M., Ahmad, S., Khan, S.A. and Sharif, A. (2020). Blockchain Technology in Agriculture and Food Supply Chain Management: A Comprehensive Review. *Computers and Electronics in Agriculture*, 179, 105820.

[7] Krishna, H.K., Anurag, S.R. and Bhuvaneshwari, V. (2018). Application of drones in agriculture: A comprehensive review. *Journal of Agriculture Engineering and Biotechnology,* 1(1), 35–43.

[8] Malapert, A., Gutjahr, O., Mollard, E. and Pantin-Sohier, G. (2020). Digitalization in Food Processing: A Review. Foods, 9(6), 806.

[9] Ministry of Agriculture & Farmers Welfare, Government of India. (2020). e-Krishi Yantra Anudan Yojana. Retrieved from https://agricoop.nic.in/schemes/e-krishi-yantra-anudan-yojana

[10] Ministry of Agriculture & Farmers Welfare, Government of India. (n.d.). Khedut Portal. Retrieved from https://ikhedut.gujarat.gov.in/

[11] Pandya, P.A. and Gontia, N.K. (2023).Development of drought severity-duration-frequency curves for identifying drought proneness in semi-arid regions, Journal of Water and Climate Change, 14(3): 824–842.

[12] Patel, P.K. and Kumar, V.H.M. (2021). 2021farmers Socio-Econimic Status And Constraints Using Social Mediafor Sustainable Agriculture Development. Gujarat Journal of Extension Education. 32(2):34–39.

[13] Patel, H. and Sharma, P. (2021). Digital Revolution in Agriculture: A Study on Indian Farming Community. International Journal of Science and Research (IJSR), 10(7), 432–436.

[14] Patel, P.K., and Mallappa, V.K.H., (2022). Predictive Factors for Farmers' Knowledge of Social Media for Sustainable Agricultural Development. Indian Journal of Extension Education, *58*(4), 55–59.

[15] Rathore, P., Singh, S. and Pandey, P. (2021). Precision Agriculture Using Drone: A Review. In Smart Agriculture and Precision Farming (pp. 15–31). Springer, Singapore.

[16] Patel, S.V., Parikh, S., Patel, S. and Patel, H., "Automated Irrigation Scheduling for Different Crop using IoT," (2022) *Sixth International Conference on I-SMAC (IoT in Social, Mobile, Analytics and Cloud) (I-SMAC)*, Dharan, Nepal, 2022, pp. 100–105, doi: 10.1109/I-SMAC55078.2022.9987433.

[17] Singh, A. and Mishra, A. (2020). Drones for Crop Monitoring and Management in Indian Agriculture. In Smart Technologies for Sustainable Smallholder Agriculture: Upscaling in Developing Countries (pp. 123–137). Springer, Singapore.

[18] Singh, A. and Pardeshi, I.L. (2019). A Review on the use of drones in agriculture. *International Journal of Recent Technology and Engineering (IJRTE),* 8(2), 7300–7305.

[19] Singh, S., Yadav, S. and Verma, S. (2019). Application of drone technology in agriculture: A review. *International Journal of Chemical Studies*, 7(5), 2918–2921.

[20] Singh, V. (2020). Digital Agriculture in India: Opportunities, Challenges, and Future Directions. *Journal of Agriculture and Environmental Sciences,* 9(2), 29–35.

[21] Singh, V., Singh, S. and Kumar, P. (2020). Role of drones in agriculture: A review. *Agricultural Reviews*, 41(3), 242–249.

[22] Swinnen, J. and Kuijpers, R. (2020). Value chains, food standards, and food security. *European Review of Agricultural Economics,* 47(5), 955–979.

[23] Zhang; sun, x, song, l; chen, w; and Zang, x.(2019) Smart irrigation syplems for sustainable Agriculture. A. Review; q(q)-183.

Smart, Sustainable and Space-Agriculture

Rajendra Shende[1]

Introduction

Agriculture is not just a food producing industry. It is the holistic interlinked value added system that deploys nature's resources to support life on the earth and nurture the safe future, in wake of the human development. Sadly, this crucial and life-supporting system has become utterly vulnerable to the climate change. The dilemma is that agriculture is also hugely responsible for the climate change due to emissions of the greenhouse gases (GHGs) that causes global warming.

Negative Impacts

The agri-sector is experiencing negative impacts from record breaking temperatures, extremely uncertain rainfall of extreme nature, invasive and alien pests. These adverse impacts of climate change are already at the door steps of farmers. The consumers are feeling the heat of the record-shattering temperatures and also because of the high prices of agricultural products. These impacts are getting more frequent and more severe in their intensity year after year. Unless the actions are taken at the unprecedented level to mitigate the adverse impacts, the food security of 8.1 billion people will be in jeopardy. The flora and fauna which are integral part of the food system are also facing existential threat.

Agriculture definitely lags behind other sectors in terms of addressing climate targets, commitments and actions. Yet it has potential to become an important part of the overall mitigation solution by reducing GHG emissions and removing CO^2 from the atmosphere by sequestering carbon.

In total agriculture, forestry and land use (AFOLU) represents around one-fifth (20–22%) of global anthropogenic GHG emissions. Half of these stems from methane and nitrous oxide (on-farm emissions), and the other half from CO^2 emissions resulting from land use, land use change and forestry (LULUCF).

---

[1] Former Director UNEP, Coordinating Lead Author of Nobel Peace Prize-winning IPCC, IIT Alumni, Founding Director of Green TERRE Foundation.

Fortunately the double opportunities to reduce both direct and indirect emissions exist. Nature-based solutions to CO_2 removal from the atmosphere, through carbon sequestration (biomass and soils) are significantly important for near term mitigation. Such mitigation will be also accompanied by enhancing the agri-productivity.

Lagging Behind in Climate Action

Despite this potential, agriculture sector does not get priority in the climate policies, actions and commitments. Countries, including India, there are no specific targets and policies for mitigation of GHGs from Agri-sector. Subsequently there are no plans for targeted subsidies and incentivisation of mitigation. Agriculture receives considerable policy support for economic aspects, however, that fails to stimulates innovations for mitigation and climate resilient agriculture. Contrarily, a large share of existing support to agricultural production in many countries, including India, potentially contributes to increasing their GHG emissions.

So What Needs to be Done in Indian Context?

Addressing the climate challenge and at the same time realising PM Modi's wish to double is possible. In short term, while mitigation efforts could be intensified in India, attention is required to be given thought Krishi Vigyan Kendra (KVK) for climate resilient practices, seed management, soil conservation, and adapting to changing climates in the various regions of India. Similar to establishing additional IITs and IIMs over last 10 years, India needs Climate Resilient Agriculture Centres (CRACs) to 'crack' the agricultural dilemma.

The existing 70 Agri-universities across India could act as catalytic knowledge centres and pilot units to support CRACs for mitigation efforts and climate resilient practices. The campuses of these universities, along with the students and faculties, are the living-laboratories for piloting innovations in Agricultural sector for mitigation and adaptation.

Network of Universities and HEIs

Green TERRE Foundation, a not-for-profit organisation under sector 8 of company act of Ministry of Corporate-affairs has initiated the network of universities to promote the practical pilot activities in the campus to contribute towards United Nations Sustainable Development Goals (SDGs). The network is called Smart Campus Cloud Network (see:

sccnhub.com). It is the global network of the educational campuses of universities or colleges. It is Green Skilling movement that adopts 'learning by doing' and 'accelerating by sharing' practice to scale-up and speed-up the much needed transformation.

The network in India is supported by UNESCO, Ministry of Power, Ministry of Education, AICTE, ASSOCHAM, UGC and NITI Aayog. 500+universities and Higher Educational Institutes are registered with Green TERRE Foundation.

The sub-set of this network is Agri-Universities to address the triple challenge: Mitigation of emissions from Agri-sector, Adapting the climate resilient practices and doubling the farmers income. All the four Agri Universities in Maharashtra are the members of the network. The agri-universities in Gujarat are being contacted to join the network. ICAR is also contacted to guide this network.

What is the Purpose of the Network of Agri Universities?

The network, primarily, engages the students and the faculty to carry out practical and SDGs-friendly and Climate-friendly activities within the campus. The youth in universities are the change agents, and future climate-champions. They can re-imagine the agriculture and food-supply system. Campus is incubator of creative practices. The campus is also where minds of the students are moulded and actions are accelerated. Once the piloting results are scaled-up and once KVKs spread the practice and innovations, India can then be the leader in addressing climate change.

Suggested piloting of the projects in cooperation with other campus are:

1. *Smart Agri:* Deploying the digital technologies like AI, IoT, Data analysis, blockchain to ensure sustainability of Agricultural food chain, optimisation of water use, providing the optimum micro-nutrients to soil.
2. *Sustainable Agri:* Deploying Solar-panels for electricity in the arid farms for
   (a) plantations of heat sensitive plants under the panels,
   (b) water collection from dews/moisture on the solar panels,
   (c) using the places under the solar panels for cattle-care, stable and poultry.
3. *Natural Farming,* organic farming, and agroforestry
4. *Bioenergy* from biogas with catalysis and agri-residue.

5. *Renting sheep* for lawn mowing in the campus.
6. *Renting bee-hives* for enhanced pollination,
7. *Developing Seed Banks* for climate resilient species, varieties.
8. Space Seed Breeding with help of ISRO to benefit from Cosmic Rays and microgravity which is the latest movement in countries like USA and China to ensure food-security and climate-friendly practices.
10. *The mitigation of methane and nitrous oxide* emissions from rice plantation.
11. *Policy Research* for climate resilient crops like millets, optimising the water use in rice, cotton and sugar cane plantation.

Ancient Indian Knowledge on agriculture was carefully harvested by the Indian sages over thousands of years. Careful observations of Earth's cyclical ecosystems, mindful studies of nature and scrupulous analysis of flora and fauna has resulted in practicing of the sustainable agriculture. The seminal principle of 'living **with** nature' and not 'living **on** nature' is part of Indian Knowledge System.

Five basic elements that make the nature, as recorded and described in ancient scriptures are Earth (Prithvi), Water (Jal), Fire (Agni or Tej), Air (Vayu) and Ether or Space (Akash).

Interestingly, we have deployed all the elements except Akash (space) for Agricultural practices.

I feel that we can now leverage Akash (space) through Space Seed Breeding Programme to enhance the productivity when those seeds are brought back on the Erath. University campuses are the breeding grounds for such innovations.

Transformative Agricultural Extension Strategies for Future: Policy Implications and a Way Forward

V.V. Sadamate[1]

Introduction

Farm Sector: Impressive Performance

The agriculture and allied sectors in India experienced impressive growth in the past two years, registering 18.8 per cent contribution in GVA (2021–22) and a growth of 3.6 percent (2020–21) and 3.9 per cent (2021–22). The sectoral production performance is impressive: food grains: 316.05 mt, horticulture production: 331.05 mt, largest producer of milk 210 mt, meat 8.80 mt, eggs: 122.11 billion eggs per annum and fish production:14.5 mt.). Extension Services, both public and private have played crucial role in technology delivery.

Agricultural Extension Models in Operation

Agricultural Extension Models(both public & private) evolved over a period of time would include:(i)Research Institutions Outlets like Krishi Vigyan Kendra (KVKs), (ii) Development Department outlets, (iii) Commodity Board Outlets (Spices, Coconut, NDDB, NFDB, Tea, Coffee, etc., (iv) Private Sector Interventions: Entrepreneurs, PPPs, Start-ups, CSR provisions, Input agencies, (v) Financial Institution Interventions like NABARD, RRBs, DCBs, etc., (vi) Farmers & Farmer Organizations, (vii) NGOs: Local, Regional, National & International, (viii) Media (Print, Electronic), (ix) Social Media and (x) Externally Aided Projects and Donor Driven Reforms Models.

Major Extension Constraints

Several Extension constraints are observed over different extension models considerably affecting the technology dissemination process. The crucial

[1] Agricultural Extension Specialist & Former Adviser Agriculture, Planning Commission (Now NITI Aayog), New Delhi, India. E-mail: sadamatevv@gmail.com

ones include: (i) Policy support and systemic inadequacies, (ii) inadequate investments and extension infrastructure, (iii) gaps in need based deployment of manpower at various levels, (iv) weak extension set up in allied sectors like horticulture, livestock, fisheries, agro forestry and speciality agriculture, etc. (v) outreach constraints of existing extension models, (vi) inadequate capacities of the district/block level extension agencies to respond/address the field problems, (vii) need for convergence of extension efforts, (viii) need for private paid extension efforts, (ix) need for farm youth and farm women specific strategies, (x) scope for R&E linkages and feedback management, (xi) enhancing ICT application, (xii) need for value chain management, greater market integration and business orientation, (xiii) technology specific and location specific requirements, (xiv) skilling farmers, field functionaries and stakeholders, (xv) intensive research required in extension systems, (xvi) need for inbuilt M&E for timely correctives in extension actions, and (xvii) need for capturing international experiences for strategic advantages, etc. These constraints could be addressed through systematic policies, programmes, investments and operational interventions at various levels.

Agricultural Extension Outreach is a Huge Task

Further, Extension Delivery is a huge task to cover over 700 districts, around 7000 blocks and nearly 7 lakh villages covering 127 agro-climatic zones & 14 Crore Farmers. Nearly 83% of them are small and marginal ones. The farming population is spread over varied social dimensions, varied geographical situations, varied resource conditions. Therefore, agricultural extension is a tough articulation and architecture to make a significant impact.

Urgent Need for Extension Innovations/Reforms for Future: Crucial Ones are Detailed

The delivery approaches, strategies and models need to get transformed to deal with the future outreach challenges in agricultural and allied sectors. The crucial ones are dealt in brief as follows.

Awareness Campaigns on the Scheme Provisions & Availing them

Awareness campaigns are needed especially at the block levels through agriculture and allied sector departments, KVKs, NABARD, etc. Involving FPOs, NGOs, CSR partners, PRIs, Cooperatives, SHGs, etc with focus on availing provisions available under various schemes and modalities to access the same. "Krishi Mohostava" model adopted successfully by the State of Gujarat is classical example. The Senior as also the filed functionaries move all through the blocks and clusters and share programmes of the Central, State Governments as also of the other agencies across the sectors. The media back up (print, electronic and social) and collation of feedback form the field for appropriate actions are the significant features. Next decade should focus on this very crucial aspect actively to provide access to large number of beneficiaries.

Empowering Farmer Aggregates for Managing Extension Services at the Cutting-Edge Level

Promote farmer aggregates like SHGs, FIGs, CIGs, FOs, FPCs, Cooperatives, etc., and enhance their skills for better production, marketing, and price negotiating/programme delivery capacities. Social capital formation, community led initiatives, participatory approaches, involvement of community resources need to be encouraged in the extension processes. Farmer Organizations (FOs)/Farmer Producer Companies (FPCs) should come up on a cluster basis with the strong market integration, where they may act as a single window service centres to all farmers/producers in a given production cluster. The small, marginal and weaker sections of the farming communities may be duly represented and sustained in such aggregations.

Enhanced Outreach of the Extension Models in Operation

It is important to improve outreach, interplay and performance of extension models to their optimum levels by (i) widening the sectoral and area coverage, (ii) partnerships and resource sharing, (iii) enhancing integrated delivery, (iv) improving penetration of services to the small producers, etc. There is need to promote chains of extension agents, formal/informal, such as Para techs, start-ups, ICT platforms, entrepreneurs, ACABCs across the production systems and build their capacities for effective extension delivery. Market-led extension strategies are required to be worked out involving farmer aggregates. Further, increased investment in market infrastructure (electricity, warehousing, rural roads, modern ICT etc.) by the government, private sector and local communities are urgently spelt and operationalized.

District/Area Level Farm Science Centres, like Krishi Vigyan Kendras (KVKs- Farm Science Centres - India) mechanism may be promoted and strengthened for technology assessment/refinement and demonstration of the frontline technologies to respond to the emerging needs of the farmers. Eventually, the KVKs may stand up as strong knowledge resource base for public and private extension services and across the production systems.

Extension for Allied Sectors is Weak: Needs Urgent Addressal

In allied sectors like horticulture, animal husbandry, dairy, poultry, fisheries, nutri-cereals, etc., the extension outreach is very weak, although these sectors contribute significantly. Hence, extension needs to be re-organized in allied areas by way of: i) providing additional subject matter specialists (SMS) to the district/block levels in potential areas, ii) promoting growers' associations, commodity groups, entrepreneurs, iii) promoting dairy/fish cooperatives and using dairy/fisheries entrepreneurs as extension agents. Extension for disadvantaged areas/groups could be addressed through innovative extension solutions like i) combination of extension service provider (ESP), ii) involving NGOs/Para techs, iii) intensive use of print/electronic/mass media/ICTs/Social media/Folk media, and iv) participation of local communities and institutions, etc.

Farmers Field Schools to Farmers Business Schools

Farmers Field Schools (FFSs)/seed villages proved as an effective extension tool, practiced widely, needs to be up-scaled, and not only for crops but also in the allied sectors too, as these provided opportunity for farmer-to-farmer knowledge dissemination and capacity development, promoting farmer led innovations and farmer first extension. National Farmer's Innovation Fund could be established to promote farmer to farmer extension. Farmer first approach needs to be promoted with focus on critical needs of the farmers in a given micro ecology and work out options for their redressal. Farmer Business Schools tried by FAO in Africa, South East and Latin America may be tried and promoted.

Addressing Extension Needs of Farm Women and Farm Youth

It is important to create enabling environment for women extension personnel in extension services. Enhanced access to credit/inputs for farm women and providing gender sensitive and home scale nutritional extension services are crucial. Specific measures proposed include sensitizing/reorienting the extension services on gender and nutrition issues, developing and piloting such extension models, promoting nutrigardens and alternate nutri-foods, women empowerment, enhanced use of digital networks, public awareness and stakeholder participation. Pilot studies in vulnerable areas and capacity building at various levels is strongly recommended.

There should be adequate focus on motivating and attracting youth as entrepreneurs in agriculture for which high-tech agriculture, secondary agriculture, processing and value-addition opportunities, innovative marketing, commercial enterprises, etc. should be targeted. Youth in agriculture needs to be provided institutional and financial support through R&D organizations, agri-clinics, financial institutions, etc. for promoting them as development agents/job providers. Committee of Food Security (FAO High Level Panel of Experts-2021) suggested youth engagement strategies like: securing dignified and rewarding livelihoods, social recognition, increasing equity and rights to resources, enhancing knowledge, education and skills, fostering sustainable innovation. Incubation programmes and start-ups be promoted involving youths.

Integrating Private Sector Extension Services and Efforts

Public-Private Partnerships (PPPs) are required down the line to promote MOU based participatory extension arrangements. Public sector would need to be oriented on contribution and strengthening of PPPs. The capacity of private agents also needs to be improved by strategic alliances which need policy directions. Agri-clinics and Agri-business Centres (ACABCs), agri-entrepreneurs, agri start-ups, e-platforms, etc. are found to be gaining space in extension operations. Agribusiness MBAs, IT graduates and farm youths (including school dropout, men and women) may be promoted for providing advisory services on payment basis in collaboration with public and private sectors. There is need to promote Private Paid Extension (PPE) services in commercial/horticultural crops as practiced in developed countries and in agriculturally well performing States.

Strengthening Research and Extension Linkages and Feedback Management

There are various research and extension (R&E) linkage forums available in agriculture and allied sectors at various levels (national, regional, state, district and even at the block level). The performance of these linkage forums needs a lot of improvement in terms of contents, coverage, joint actions and follow-up. Technology options for agro-climatic zones and sub-zones needs to be systematically worked out as per socio-economic dimensions of each micro agro-eco situation. This would enhance dynamism in extension processes. SMSs of the KVK/SAU System should mentor the block level formations of the development departments. Farmer—Scientists Interactions (FSI) may be upscaled in knowledge driven technologies. Likewise, successfully demonstrated innovations by the research agencies should be up scaled by the Field Extension agencies. Greater involvement of the scientists is needed in technology driven extension like NRM, IPM, INM, conservation agriculture, farm mechanization, climate resilience, etc. Knowledge-driven innovations – extension agents may first work some time with scientists and then with the farmers for a considerable time, show the results through demonstrations and then upscale innovations, preferably through the farmer aggregates for wider footprints.

Feedback on collation of information from the farmers and field functionaries is seemed to be weak in present day extension system, so also its systematic documentation and analysis. The collation of feedback (both positive and otherwise) from the farmers and field functionaries may be the first essential step, followed by documentation/analysis at the district/block level. The review and reporting could be taken up by the regional set up. Such an arrangement would not only make the extension operations demand driven but may also facilitate providing policy signals to the higher-level R&E management. Feedback management mechanisms are pilot tested across the sectors and for micro agro situations then learning up scaled.

Streamline Training and Capacity Building of the Field Agencies and Farmers

National/State and District Training frame work be suitably strengthened in terms of manpower and infrastructure. Interplay of these institutions needs to be worked out systematically, sharing the output-input relationships and the experiences. Farmer's training in agriculture and allied sectors is shared by the large number of agencies/organizations at the district/block levels. It is suggested that focused and segregated training responsibility needs to be assigned to different agencies as per expertise and competence of the agency. Innovative farmers, credible farmer cooperatives and FPOs, successful Agri Clinics, effective IT platforms could be integrated in the training and CB strategies as applicable in agriculture and allied sectors and as per demands and special felt needs of each categories of farmers, farm women and farm youth.

With the change in present day farming scenario, farmers' skills need to be up graded/oriented towards entrepreneurship, income/business orientation, value added/secondary agriculture, aggregated production and marketing, etc. Accordingly, extension workers' skills need emphasis on social skills, entrepreneurial management skills, technical skills, and media management skills. Skill enhancement of the farmers, field functionaries and SMSs (subject matter specialists) need to be prioritized to make them ready for area-specific, commodity specific, technology-specific and eco-region-specific extension as applicable to the micro agro situations and vulnerable groups.

Priority Setting and Convergence of Extension Efforts at the Block/Cluster Levels

There are series of district level planning instruments, scheme-wise. The extension functionaries may scan these instruments for capturing the right priorities across the schemes/programmes. The crucial extension issues must reflect in the systematically developed Block Extension Plans (BEPs) at the cutting-edge levels, defining the role of various extension service providers (ESPs). Well established/networked R&D platforms at the cutting-edge level could be a convergence arrangement. The resources from the converging departments would need to be pooled and the roles and responsibilities required to be delineated carefully harmonizing of work plans of the related Federal/State schemes. Convergence requires proper role space and resources amongst stakeholders for mutually agreed Block Action Plan (BAP). Matrix mode approach is required indicating the ongoing programs and the gaps to be bridged. Operational flexibilities to the converging partners need to be worked out in advance and provided for. Large number of convergence pilots be carried out both for resource poor and resource endowed areas and learning inculcated in the ongoing programmes and schemes.

ICT and Media Management to Accelerate the Extension Delivery Processes

Social Media (WhatsApp, Facebook, Twitter, Instagram, emails, blogs, vlogs, App-based services, internet platforms, block chains, you tube channel, etc.) are powerful communication tools that enhance the beneficiary coverage in the shortest period of time which could be used effectively in networking farmers' and offering context-specific information. Further, dedicated tele-numbers could be provided by the R&D agencies like Kisan Call Centres (Farmers' Call Centres) in India that provide country wide common eleven-digit toll free number 1800-180-1551 for providing tele replies to the farmers' queries in local languages. ICTs are also being used to strengthen the capacity of extension officers and field staff to first reach the farmers with timely and accurate information and help capture data from the field. The e-Choupal initiative of ITC Corporate, as an example has had positive effect, the system enhanced supply chain efficiency.

Advanced high techs like drones, satellite systems, sensors, artificial intelligence, machine learning, robotics, data analytics, etc. are used for precision farming, improved farm management, providing real-time solutions, etc. E-platforms and mobile applications are emerging very fast linking farmers on one hand and markets and PVA chains on other. However, the pertinent question is how to increase its access to the large number of farmers, reliability of the information and community demos. Therefore, need for enhancing capacities of the service providers and promoting farmer/farm women knowledge groups (FKGs/FWKGs) at the grass roots level to capture and adopt the high-tech interventions. This approach should be the game changer for future extension strategies and operations in developing countries in Asian and African regions. The R&D outlets of the Farm Universities and the private organizations at the District/block levels should be capacitated to demonstrate impacts of such technologies. Successful demonstrations and experiences should be widely replicated.

Re-Orient Communication Strategies

Use of media combos and supportive extension methods be intensively used in disadvantaged and far-flung areas. Such combinations should be worked out as per subject matter, agro-climatic and socio-economic criteria of the farmers. Location specific Farm Telecasts, Radio broadcasts and Community Radio Stations (CRS) should continue to play significant role in farm information dissemination process. Important considerations for farm radio/TV broadcasts are: content development, treatment, delivery mode and the real time impact in the field. The Directorates of Extension of the Agri - Universities and farm information wings of the State Departments may need to revisit their role when the private players and IT platforms are proactive now.

Research in Extension—Revisit to Make it Demand Driven

Research in Extension is a crucial area but not attended adequately. Strong extension research input is required from various academic and semi academic stakeholders to evolve a body of extension knowledge and strengthen it. Farmers, their field endowment, socio-economic setting, innovations/technologies, adoption process are to be looked in to seriously. Farm sectoral variations, are the research variables for overall impact of the extension interventions. The extension research labs need to be promoted in competent organizations and at various levels, for examples the Centres of Excellence of the MANAGE in India could take a lead in this process.

The future extension research strategies must draw strengths from international, private sector, NGOs and multilateral donor experiences. Future extension research may focus on systems interplay, convergence, agribusinesses and entrepreneurs, App based ICTs, extension for unreached, climate change adaptation, etc. Extension Research outcomes be ploughed back for reforming existing policies and operations. Suitable state specific mechanisms may be worked out by the SAMETIs, ATARIs and the Directorates of Extension of the SAUs.

Interventions/Technologies that Need Intensive Extension Efforts

There are good number of farm innovations & technologies that need serious attention of the development agencies for wider dissemination through effective extension strategies and cost reduction applications. Some of the crucial ones include: (i) Utilizing barren lands as solar energy hubs, (ii) Roof top solar panels – promotion in rural areas, (iii) Promoting Climate Resilient Agriculture and Climate Smart Farmers with involvement of farming community at the grass roots level, (iv) Soil health management, especially nutrient application as per the recommendations of soil health cards,(v) Promotion of one district one product scheme for export opportunities, (vi) Greater integrated developmental focus on the identified aspirational districts/blocks, (vii) Drip irrigation and protected cultivation. (viii) Tissue culture technology produced planting material, especially in horticultural crops, (ix) Raised bed cultivation and zero tillage technologies, (x) Organic Agriculture and Natural Farming, (xi) Popularization of Green Carbon Credit Scheme (xii) Small scale farm mechanization for hilly and mountainous areas and horticultural crops, (xiii) Application of drones, sensors, artificial intelligence, robotics, machine learnings, etc.,(xiv) Feed and fodder technologies, (xv) Health management practices for the small ruminants & poultry, etc.

Farm Youth as Extension Entrepreneurs

Further, the youths are looking for high-tech agricultural options as opposed to the traditional agriculture. There is another segment of technical and management graduates looking for entrepreneurial and value chain opportunities in farm operation linking farmers to the best of the technologies on one hand and to the best of market opportunities on other. Large number of start-ups and private sector initiatives as also empowered farmer organizations and farmer producer companies are coming up in agriculture and allied sectors. This is a very positive side of farming wherein high-tech production and marketing interventions are integrated. Such initiatives are also looking for market/export opportunities abroad. States like Maharashtra, Gujarat, Karnataka, etc. are promoting such agri entrepreneur and farmer driven initiatives. In Maharashtra, the Farmer Organizations are federated at the district and State levels, drawing strengths by mobilizing the producers and marketing operations.

Private Sector Led Extension Efforts

Horticulture segment is a classic example of these initiatives wherein the public extension services are generally found to be weak. However, the private interventions have come up in a big way through start-ups, entrepreneurs, partnerships, value chain operators, planting material producers, aggregators, processors (public- private), etc. Apart from fruits, the private initiatives have grown significantly in the vegetable, floriculture, apiculture, mushroom production, etc. It can be confidently stated that the horticultural extension is primarily private sector driven as in case of poultry or fisheries. The fisheries and animal husbandry extension services are also looking up for big private sector push in the days to come. The face of public sector extension has been changing in the present context as enabler and promoter of private players through schematic back up in the relevant sectors.

Supportive Enablers for Efficient Extension Performance

Farming System Led Extension Strategies

It is essential that extension functionaries should demonstrate and emphasis to the farmers adopting the "Farming Systems Approach" such as grain crops on priority, while as per the situations and environmental conditions livestock and fisheries be promoted as another essential major supplemental nutritive food sources. Likewise, horticulture and vegetables also be prioritized as per micro-agro ecological potential. The above combination (grain crops, livestock and fruits and vegetables) may prove to be the most effective system and win-win situation against various climatic risks and calamities which is need of the day for the farmers in developing countries, desperately trying to attain their food security.

Benchmarks for Extension Performance

Benchmarks for extension performance may now need to include extent of adoption of IFS approach, extent of programmatic convergence with other departments, enhancement of production/income of the producers, enterprise combination, crops and cropping pattern diversification for the economic advantage of the farmers, market linkages, etc. Extension accountability needs to be systematically designed, cultivated and promoted through farmer aggregates.

Processing and value addition are essential to add to the farmer's income whenever there is marketable surplus particularly when the perishable produce is more and is in excess to the demand. It has a great proven promise in revisiting the same and replicating.

Extension Delivery is a Complex System—Needs Perfect Understanding on Socio-Economic and Technology Dimensions

The Extension delivery is a very complex system dealing with farm technologies on one hand and the socio-economic dimensions on the other. In a given micro-situation in the absence of reformed extension services, the programme delivery to the targeted clientele is inadequate and of a weaker impact in the field. Hence, strong extension innovations/organizational and management reforms are pleaded both in public and private extension systems and in the delivery mechanisms. Extension innovations and reforms on some needs to be strengthened and a few others pilots tested to suit the local needs and resources.

There is need for greater convergence and coordination at various levels. Needless to say, that the extension service delivery would be far more visible, efficient and location specific and most productive if it is suitably backed by the public policies, investments, incentive linked good agricultural practices (GAPs), market reforms, strategic scaling innovations and credit – input augmentation.

Indian Agricultural Extension Experiences—Wider Replication

Indian Extension experiences are rich in content and delivery. The emergence of pluralistic models both on frontline and field extension side, suiting the macro and micro agro climatic need to provide much needed flexibilities in extension dynamics. Each State has unique extension strategies developed on public and private interventions across the production systems. Some of the sectors like horticulture, poultry, fisheries etc., are driven more by the private sector extension services making significant differences in the productivity and income of the farmers. The extension models driven by the scientist (frontline extension), the ones carried out by the development departmental efforts (field extension) and the efforts of the private sector extension are showing varied impacts. India therefore could play an important role in capturing best extension practices and their dissemination to the developing countries, especially in South Asia.

Need for Regional Extension Alliances

For effective exchange of learning and experiences, there is need to promote Regional Extension Alliances for sharing innovations, experiences, cross learnings etc. Recently promoted Agricultural Extension Platform for South Asia (AEPSA) is an appreciable initiative in this direction. This would enable scouting transformative innovations, scaling strategies, joint programming, joint project planning and implementation, sharing evaluation experiences, faculty and student exchanges, joint dialogues, and exposure visits to capture the good extension practices. Serious efforts are needed to fund such extension efforts by the national and international agencies.

Unlocking Potential: Sustainable Development in Arid Areas

Suresh Acharya[1]

Introduction

India's arid zones span approximately 35 million hectares across seven states, with Rajasthan and Gujarat contributing over 80% of this area. Despite facing numerous challenges, these regions also offer significant potential for sustainable development. Arid zones are characterized by low and variable rainfall, high winds and intense sunshine, making them the most fragile ecosystems where minor disturbances have lasting impacts. India's arid regions are concentrated between 24° and 29°N latitude and 70° and 76°E longitude, with the Thar Desert dominating the landscape.

In arid areas, the primary challenge lies in the scarcity of freshwater. Traditional irrigation methods often deplete groundwater reserves, while overgrazing, deforestation and improper land management practices worsen soil erosion and desertification. Climate change has further exacerbated these issues by increasing evaporation rates and disrupting ecosystems. As a result, communities in these regions have historically faced acute water shortages, leading to significant migration in chase of water sources. However, with recent technological interventions and government schemes, water management has taken precedence over simply searching for water in these regions.

A multifaceted approach is pivotal for managing challenges in arid regions. Programmes like the Desert Development Programme (DDP), launched in 1977, focus on watershed management, rainwater harvesting and promoting drought-resistant crops to improve water security and land fertility. The Integrated Watershed Management Programme (IWMP) takes a holistic approach, emphasizing community participation and water conservation techniques like laser land levelling and micro-irrigation. Beyond water, the *Deen Dayal Upadhyaya Grameen Kaushalya Yojana* (DDU-GKY) equips rural youth with skills to create alternative income sources and reduce dependence on agriculture. Renewable energy

---

[1] AGROCEL Industries Private Limited, Field R&D, Koday, Kachchh, Gujarat, India.

initiatives like large-scale solar plants aim to decrease reliance on fossil fuels. Finally, investments in precision agriculture, desalination technologies and drought-resistant crops demonstrate a commitment to innovation for addressing the unique challenges faced by arid regions. These initiatives work together to transform arid landscapes into thriving, sustainable ecosystems that contribute to India's overall development.

Understanding Arid Regions: A Data-Driven Approach

Managing arid areas effectively hinges on a deep understanding of their land, water resources and the impact of implemented solutions. Fortunately, advancements in critical and emerging technologies offer powerful tools for data collection and monitoring. Remote sensing satellites provide valuable insights into vast landscapes, revealing land cover, soil moisture and vegetation health. Geographic Information Systems (GIS) act as a central hub, integrating various data sources to create a comprehensive picture of resources, infrastructure and development interventions. Additionally, sensor networks deployed across fields and water sources provide real-time data on soil conditions, water levels and weather patterns, empowering informed decision-making for sustainable resource management.

Optimizing Water Use Efficiency

Arid regions present a unique set of challenges for development due to limited water resources, harsh climates and fragile ecosystems. However, advancements in technology offer promising solutions to overcome these obstacles and promote sustainable growth in these areas. Desalination plants, particularly those using reverse osmosis or membrane distillation, are turning saltwater into freshwater, providing a vital new source. Advancements in energy efficiency and integrating renewable energy are making this process more sustainable.

Wastewater is not waste anymore. Advanced treatment technologies are transforming it into irrigation-ready water, easing the pressure on limited freshwater supplies. Precision irrigation techniques like drip irrigation and laser land levelling deliver water directly to the plants, minimizing waste. For areas shrouded in mist, fog harvesting nets capture precious moisture droplets, offering an additional freshwater source.

Beyond individual technologies, closed-loop systems like aquaponics and hydroponics use minimal water compared to traditional methods.

Aquaponics even creates a win-win situation, combining fish farming with plant cultivation in a water-efficient manner. At the plant level, advanced tools like AI, nanotechnology and tensiometers help create precise water budgets for each crop, considering weather conditions, soil types and growth stages. This ensures every drop count, especially during the crucial, limited "least growth period" in arid regions, where early maturing crops are a must.

Water efficiency is not just about the plants. Sensors, drones, AI and data analytics enable precise water budgeting from local ponds all the way up to river basin levels. This focus on water use efficiency and irrigation optimization works hand-in-hand with other strategies like crop diversification, early warning systems and fair water pricing to create a holistic approach. Satellite imaging, artificial intelligence and drones provide real-time data on water levels and soil conditions.

Technologies are being developed to address sediment management, preserve biodiversity and reduce the carbon footprint of water infrastructure. By embracing these advancements, arid regions can turn the tide on water scarcity and create a thriving future.

Water Management Initiative in Arid Kachchh, Gujarat

The initiative in Kachchh, Gujarat, focuses on participatory innovative water management strategies to address severe water scarcity and degraded water quality in arid areas. This includes river basin management, deepening ponds and using anti-transpiration tools. The Rukmavati River basin management initiative showcases the transformative impact of water harvesting on agriculture in the region, demonstrating positive outcomes in an area facing severe water scarcity and degraded water quality.

AGROCEL Industries Private Limited, as part of its Corporate Social Responsibility (CSR) efforts, undertakes several water management initiatives. In 2023, it deepened 20 ponds with a total capacity of 125,416 CuM. Additionally, it also implemented Individual Rainwater Harvesting Structures (RWHS) at 139 units in harvesting 1175.24 CuM of water. Furthermore, 8 Institutional RWHS projects were executed, collectively harvesting 263.00 CuM of water. As part of the Kaka Centenary Programme on Jal Mandir, AGROCEL has successfully established 216 Jal Mandir structures, with an estimated total water harvesting capacity of 8,76,660 CuM. Efficient irrigation techniques like drip, sprinkler and laser irrigation are promoted to optimize water utilization, minimize losses and mitigate soil salinization risks. The initiative also embraces the 3Rs (Reduce, Recycle, Reuse) by repurposing wastewater for irrigation, using innovative technologies to enhance precision water use and strengthen resilience against adverse weather conditions.

Anti-evaporation techniques in water-resilient agriculture play a crucial role in minimizing water loss from open water bodies in arid areas. These methods include floating covers, monolayer films and the use of chemical additives to reduce surface tension and evaporation rates. Planting windbreaks around water bodies serves as a natural barrier against wind-induced turbulence. An experiment conducted by AGROCEL Industries Pvt. Limited in Ludiya Village, Gujarat, showcased the effectiveness of Evaloc, a fatty acid molecule, in reducing water evaporation. The experiment showed a 35% reduction in water evaporation when Evaloc was applied alone and a 43% reduction when combined with polythene. This innovative approach saved approximately 42 lakh liters of water, meeting the village's needs for about four months at a cost of Rs. 40,000, with a per liter water-saving cost of approximately Rs. 0.95.

Community participation is key, with the initiative engaging farmers and local communities in decision-making processes regarding water allocation and infrastructure development. Demand-side Water Budgets are prepared to ensure optimal utilization of limited water resources. AGROCEL's holistic approach to water management serves as a model for sustainable water utilization in arid regions, emphasizing water conservation, rainwater harvesting and efficient irrigation techniques.

Agroforestry

Integrating trees or shrubs into agriculture diversifies farming systems, providing multiple income streams. Among the different initiatives for the development of arid areas, enhanced biodiversity featuring 35 different medicinal species and Cacti species with medicinal and fodder value were also used. Agro-forestry efforts focused on screening multi-utility trees, shortlisting *Simarouba* and *Mangium* for further scaling up. *Simarouba*, a fast-growing light timber tree, provides high-quality edible oil and litter for soil improvement. *Mangium*, known for its valuable timber, showed resilience under saline and arid conditions. Noni, a wellness plant, also thrived in such environments. Local biodiversity conservation included planting 170 variable genotypes of *Sonamukhi* (*Cassia angustifolia*) and maintaining variants of local trees. Additionally, various tree species were planted in the biodiversity park, such as neem (*Azadirachta indica* A), babul

(*Acacia nilotica*), herma (*Acacia leucophloea*), khijdo (*Prosopis cineraria*), seven (*Gmelina arborea*), thor (*Euphorbia spp*), karanj (*Ailanthus excelsa*), jamun (*Syzygiumcumini*), kumut (*Acacia senegal*), and kher (*Acacia chundra*).The arid saline areas have unique grass biodiversity, particularly the Kedko grass, which showed exceptional performance under high water salinity conditions.

AGROCEL's "Greening the Earth" Initiative in Kachchh

The AGROCEL "Greening the Earth" initiative in Kachchh, Gujarat, demonstrates a comprehensive approach to sustainable development in arid regions, focusing on biodiversity conservation, landscape restoration and community engagement. Despite the extreme conditions in the region, such as high salinity, aridity and habitat loss, the project has implemented targeted interventions across diverse terrain types to revitalize the land and its ecosystems.

The initiative has successfully addressed four distinct project areas, each presenting unique obstacles but showcasing the potential for transformation. In the hilly terrain of Hill-1 and Hill-2, Kukma, Bhuj, the initiative encountered overwhelming odds, including lack of water and severe erosion, to establish thriving vegetation with over 18,000 and 100,000 plants, respectively, of 31 different species. The Dorek Doe site, a remote saline desert, was transformed into a green oasis through innovative water harvesting techniques. It also entailed introduction of high-quality date palms and other resilient flora, achieving an impressive 80% survival rate in otherwise abysmally saline barren land. The rocky saline terrain of Ludiya was reclaimed by employing strategic water harvesting solutions and careful species selection, demonstrating the viability of turning wastelands into thriving ecosystems. It included plants like *Rakhatrohida* or Desert Teak (*Tecomellaundulata*), an endangered slow-growing tree species with significant economic and pharmaceutical properties, including hepatoprotective, antibacterial, antimicrobial, antifungal and immuno-modulatory properties, making it an important component of greening endeavors in arid areas. AGROCEL successfully pioneered the modified Miyawaki afforestation method in Bhuj to enhance urban greenery in arid areas. By establishing four Miyawaki facilities ranging from 500 to 2000 sq m, the project has created lush, dense vegetation in limited spaces, supporting soil rejuvenation and wildlife.

AGROCEL's holistic approach, combining biodiversity conservation, landscape restoration and community participation, serves as a model for transformative initiatives in arid areas. The project has addressed adverse climatic conditions through innovative water harvesting techniques and careful plant species selection, engaging the local community as active stewards of the land. This involvement has promoted wildlife conservation, supporting a diverse array of birds and animals, including jackals and mongooses, further enhancing the region's biodiversity. The initiatives thoughtfully created sunset point, medicinal park, nutritional park, meditation point, and above all walking paths within the afforestation sites have encouraged people to immerse themselves in nature and appreciate the rejuvenated ecosystems. This approach highlights the transformative power of environmental stewardship and community engagement, serving as a model for comprehensive and sustainable interventions in other arid regions.

Integrated Approach for Countering Climate Change

The escalating levels of atmospheric carbon and declining levels in soil have become pressing issues pan world including arid area. This trend is largely attributed to increased deforestation, intensive tillage and improper soil biota management. To address these challenges, integrated pilot projects were implemented.

One such pilot focused on zero tillage and biochar application, demonstrating significant increases in soil carbon content. Zero tillage resulted in a 235% increase (reaching 1526 t/ha) compared to 58% (702 t/ha) and 67% (798 t/ha) increases with deep and conventional tillage, respectively. This means that zero tillage sequestered 3.34 times more organic carbon compared to initial levels, whereas deep and conventional tillage only sequestered 1.53 and 1.74 times more, respectively. With Biochar application initial soil carbon contentwas managed to 0.5% to 0.7% in controlled plots.

In a cotton-wheat rotation, the highest wheat yield (4035 kg/ha) was achieved by increasing soil carbon to 0.5% using 5 tons of Rich Cel/ha. With a 9.9% increase over normal carbon content (0.3%). In cotton, significant differences were observed only for biochar levels, with 0.5% and 0.7% soil carbon resulting in yields 30.5% and 29.4% higher, respectively, than no biochar application (3567 kg/ha).

Additionally, vermitea was used to enhance soil biota. Molecular analysis (BLAST) of the bacterial community extracted from vermitearevealed a diverse microbial microbial diversity, among which twelve types of microbes viz; *Terrimonas terrae, Novosphingobiumclariflavum, Bacillus oryzaecorticis, Bacillus flexus, Bacillus haikouensis, Bacillus paralicheniformis, Algoriphagusolei, Bacillus aquimaris, Acinetobacter lwoffii, Hydrogenophaga flava, Aeromonas taiwanensis* and *Bacillus firmus* were predominantly pervasive in all the Vermitea harvested from different models. Notably, only *Bacillus* and *Acinetobacter* are well-known for their beneficial roles in agriculture. Therefore, further research is needed to understand the contributions of other genera in the soil-plant microcosm.

Integrated Farming System

Diversification of systems, materials, technologies and package of practices assume significance in arid areas. Integrated Farming System (IFS) is important for assuring regular daily farm income and health of 7Js viz; *Jal* (water), *Jameen* (soil), *Jivanu* (microbes), *Jalvayu* (environment), *Jan* (people including farmer), *Janvar* (livestock) and *Jivat* (diseases & pests). An IFS model was developed for small farmers having one-hectare land. It comprised Fodder Section (0.06 ha); 3-Tier Nutrition Section entailing fruits and vegetables on 0.20 ha; Crop Section (0.70 ha), Dairy Section, Water Harvesting/Well Recharging, boundary/bund plantation, etc. Napier grass and local grass were utilized to delineate sections and fields, providing fodder for animals. Legumes like alfalfa were planted to ensure proper nutrition for the animals, while feed from cotton and mustard cake was also available. For plant nutrition, pulses were grown, and the useful microbial load in the soil was boosted through the application of endophytes/microbes, *Jeevaamrut*, and *Beejamrut*. Bhavi-Vermiwash was also produced, and trees were planted on the boundary for carbon sequestration, interspersed with shrubs like drumsticks for recurring income. This 3-tier system (lufa, fruits like noni and host of vegetables like spinach, methi, onion, tomato, brinjal, coriander, radish, carrot, etc) was used for efficient land use.

Water was harvested in a farm pond and used to recharge the aquifer in a dead well, with rainwater being filtered through vetiver grass before entering the well. About 300 man-days/year or 0.82 man/day were found sufficient for different farm activities over one hectare in the model, suggesting that other family members could engage in off-farm activities to supplement overall income.

Drought-Resistant Crops

The development of arid regions hinges on water-resilient agriculture, which customizes water usage to specific crop needs through the principles of the 3Rs—reduction, recycling and reuse of water. Emphasizing crops like *'Shrianna'* or millets reflects a commitment to mitigating global warming in these challenging environments.

Innovations such as developing Crassulacean Acid Metabolism (CAM) varieties of groundnut, which significantly reduce water use compared to traditional photosynthesis systems, are crucial. The CAM photosynthesis process uses 160–180 molecules of water to fix one molecule of carbon, in contrast to 883 and 633 molecules in the C_3 and C_4 photosynthesis systems. Typically, 1.7 L of water is needed to fix 1 gram of carbon. Consequently, the CAM system requires only 1/5 and 1/4 of the water used in the C_3 and C_4photosynthesis systems, respectively. Three CAM varieties of groundnut each for drought (DGRMB5B, DGRMB24 & DGRMB32) and salinity Lresistance (DGRMB5, DGRMB19 & TG 37A-B) were tested at Mandvi, Kachchh, out of which two varieties viz; DGRMB5 and DGRMB19 have been registered for salinity tolerance with NBPGR, New Delhi. Interestingly, plants tend to produce offspring when under stress. As a result, CAM plants can be found among C_3 and C_4plants whenever there is a threat of extinction. There is evidence indicating that these plants transition between the original C_3/C_4 and CAM versions in response to the threat of extinction. The emergence of CAM plants under stress highlights the potential for selecting CAM types to enhance water efficiency in different crops.

These advancements underscore the potential of innovative agricultural practices to address water scarcity and enhance sustainability in arid regions like Kachchh, contributing to their overall development. Advancements in agricultural research and development are focusing on the creation of crop varieties that are more resilient to the harsh arid conditions, including lower water requirements and higher tolerance to heat and drought.

Advancements in agricultural biotechnology are offering new possibilities for arid area development. Gene editing, using tools like CRISPR, is being explored to develop crop varieties with increased drought resistance and improved water use efficiency. This could revolutionize agriculture in arid regions by creating new crop varieties that thrive in harsh conditions. While these technologies offer exciting possibilities, some hurdles need to be addressed. Firstly, technology readiness and adoption costs are crucial factors. Finally, careful social and environmental impact assessments are necessary to ensure sustainable development in arid regions.

Plant-Microbes Interaction

Even under optimal growing conditions, only 24% of the potential yield is realized, with 11% and 65% lost to biotic and abiotic stresses, respectively. The situation is particularly challenging in arid areas. Experiments with beneficial microbes like *Pseudomonas* and *Bacillus* have been effective in enhancing plant growth in harsh conditions, boosting cotton yields when tested in farmers' fields. Similarly, vermitea cocktails, rich in beneficial microbes and plant growth regulators (PGRs), have also shown positive results.

Mycorrhizal fungi play a crucial role by forming a symbiotic relationship with plant roots, creating an extensive network that acts like a superhighway for water and nutrients. Combining these fungi with plant peptides has shown promising results. AGROCEL's Maha Booster Plus combines mycorrhizae and polypeptides to boost plant growth. Mycorrhizae boostmutual transfer of nutrients and water; while polypeptides, specifically LRPPs (Lateral Root Promoting Peptides), encourage the development of fibrous roots, which translates to better nutrient and water uptake, crucial for arid environments. Experiments with Maha Booster Plus have demonstrated improved vegetative growth, increased stem thickness, earlier flowering and increased economic yield. These natural solutions offer a sustainable approach to overcoming water limitations and nutrient deficiencies, paving the way for a more productive and resilient future for agriculture in these challenging environments.

Multifaceted Potentials of Bamboo

Arid regions face significant challenges, but bamboo offers a promising avenue for development due to its rapid growth, impressive carbon sequestration capabilities and suitability for diverse applications. However, harsh arid conditions, particularly high salinity levels in soil, can hinder bamboo growth. To address this, 21 diverse bamboo species were evaluated for their tolerance to saline conditions at Mandvi, Kutch, India.

The evaluation identified several salt-tolerant bamboo species with potential for arid areas, opening doors for the development of vibrant cottage industries. Among species acquired from Bangalore, *Bambusatulda* emerged as a frontrunner. Its fast growth and suitability for applications like pole production and vegetable support structures make it ideal for local artisans. *Bambu sanutan* and *Bambusabalcooa* (Beema), with their higher biomass production, are well-suited for furniture crafting, tool making and other cottage industries requiring substantial raw material.

Focusing on overall growth and development, *Bambusabalcooa* has the greatest potential for arid cultivation, with *Thyrsostachysoliveri* and *Bambusatulda* also showing promise. The study also investigated the impact of bamboo on soil pH. Species *like Dendrocalamus asper* and *Bambusa vulgaris* (Green) displayed the most significant reduction in soil pH, indicating their potential for improving saline soils. *Bambusa vulgaris* (Yellow) and *Bambusatulda* also showed promise in this area.

The success of local regeneration efforts for these bamboo species highlights the viability of cultivating bamboo in saline conditions. This approach offers a pathway for creating diverse income sources and fostering cottage industries for local communities in arid regions. With a focus on local skills and training, communities can transform this resilient resource into a foundation for thriving cottage industries, creating new opportunities and economic empowerment in arid areas.

Local Tourism

Tourism has the potential to revitalize arid regions, offering sustainable development opportunities and economic growth. These areas, with their unique landscapes, can attract tourists seeking adventure and natural wonders.

Local involvement is key for successful arid region tourism. Engaging communities can generate income through authentic accommodations, guided tours and artisanal crafts, preserving cultural heritage and meeting tourist demands.

Beyond income, local tourism can drive infrastructure development, cultural preservation and environmental awareness. Careful planning is essential to balance tourism with conservation, infrastructure needs and community benefits.

Embracing the uniqueness of arid landscapes and empowering locals can make tourism a powerful tool for sustainable development. It can create jobs, generate income and conserve these remarkable environments. A focus on a tourism model benefiting both visitors and locals ensures a vibrant future for arid regions. An excellent example is the development of Khavda's white desert in Kachchh, showcasing how tourism can transform arid areas into thriving destinations.

Renewable Energy

Arid regions, with their abundant sunshine and strong winds, hold immense potential for renewable energy generation, providing clean, reliable, and sustainable power sources.

- ***Concentrated Solar Power:*** This is a promising technology specifically suited for arid areas. CSP utilizes mirrors to focus sunlight onto a central point, generating intense heat. This heat can then be used to produce electricity through a steam turbine, similar to traditional power plants. However, CSP also offers the ability to store this thermal energy for later use, overcoming the intermittency limitations of solar panels that only generate electricity during daylight hours. This makes CSP a dispatchable source of renewable energy, meaning electricity can be produced on demand, a significant advantage in arid regions with fluctuating power needs.
- ***Wind Turbines:*** Advancements in turbine design are further unlocking the potential of arid regions. Taller wind turbines equipped with larger blades are being developed to capture stronger winds that prevail at higher altitudes. These advancements can significantly increase wind energy generation efficiency, making wind a more viable renewable energy source in arid regions with strong wind resources. The combination of these advanced technologies – CSP for dispatchable solar power and taller wind turbines for efficient wind energy capture – creates a powerful synergy, offering a reliable and sustainable solution for powering arid regions.
- ***Hybrid Energy Systems:*** Several pilot projects have been undertaken to demonstrate the feasibility and benefits of these systems. The beauty of hybrid systems lies in their ability to overcome the intermittency of renewable energy sources like solar and wind. Solar power generation peaks during the day, while wind power can be stronger at night. This complementary mix provides a more consistent energy supply for arid regions. Hybrid systems offer reliability by combining multiple renewable sources, compensating for each other's fluctuations. Energy storage adds backup power during low renewable energy availability. This reduces reliance on fossil fuels, lowering greenhouse gas emissions. Hybrid systems are scalable and adaptable, customized to meet specific energy needs based on local conditions. India's promotion of these systems is key to unlocking renewable energy potential in arid areas for a sustainable future.

- ***Showcasing Hybrid Energy:*** The Largest Energy Park in Arid Kachchh Adani Green Energy Limited (AGEL), India's largest renewable energy (RE) company, has operationalized a 551 MW solar capacity in Khavda, Gujarat, supplying power to the national grid. This achievement was accomplished within 12 months of commencing work on the Khavda RE park, where AGEL plans to develop 30 GW of renewable energy capacity, expected to be operationalized in the next five years. The Khavda RE park is set to be the largest renewable energy installation globally.
- AGEL's efforts have transformed the challenging terrain of the Rann of Kutch into a habitable environment for its workforce, showcasing its commitment to sustainable development. The region's abundant wind and solar resources make it an ideal location for giga-scale RE development. AGEL's innovative solutions and deployment of cutting-edge technology, such as India's largest onshore wind turbine generator and waterless robotic cleaning for solar panels, set a new standard for giga-scale renewable energy projects worldwide.
- Once fully operational, the Khavda RE plant is expected to generate approximately 81 billion units of clean electricity annually, powering 16.1 million households and avoiding 58 million tons of CO_2 emissions. This milestone reinforces AGEL's position as a leader in India's renewable energy sector and aligns with its commitment to India's ambitious goals of 500 GW of renewable energy capacity by 2030 and carbon neutrality.

Minerals Technology

Arid regions face unique challenges, but beneath their dry surface lie treasure troves of minerals from everyday necessities like salt to construction staples like sand and gravel. Arid regions can also be sources of energy minerals like coal, uranium, copper and zinc. For agriculture, specialized minerals can be directly extracted from seawater, including magnesium, bromine (used in various industrial applications), and even schoenite.

Developing these resources responsibly can bring significant benefits to arid communities. Mineral extraction creates jobs, generates revenue and stimulates local businesses. Local residents can participate in mining and processing, fostering economic self-sufficiency. The revenue generated can then be invested in critical infrastructure like water treatment plants and renewable energy facilities.

Technology plays a vital role in minimizing the environmental impact of mineral extraction in arid regions. Dry processing techniques significantly reduce water consumption, while desalination in coastal areas creates freshwater for processing brines. Closed-loop water systems in processing plants ensure water reuse and minimize waste.

However, for true sustainability, mineral development needs to be part of a bigger picture. Strict environmental regulations and responsible mining practices are crucial. Local communities must be involved in decision-making to ensure development plans address their needs. Education, skills training, and social safety nets empower local communities and build resilience.

Unleashing the true potential of arid regions requires collaboration. Public-private partnerships ensure responsible mining and infrastructure development. Investment in research institutions fosters the development of sustainable and efficient mineral extraction technologies. Finally, innovative financing mechanisms attract investments in infrastructure and community development projects.

By adopting a holistic approach that balances responsible resource management, community well-being and environmental protection, mineral technology can be a powerful driver of sustainable development in arid regions, paving the way for a brighter future.

Local Communities

The key to unlocking potential of arid regions lies not in external interventions, but in empowering the local communities who have thrived there for generations. Local communities possess a wealth of experience and knowledge about their environment. They understand subtle weather patterns, effective water harvesting techniques and which plants flourish in their specific soil conditions. This irreplaceable knowledge base serves as a cornerstone for sustainable development strategies. By incorporating their needs and insights into project planning and implementation, the specific need-based solutions can be found to the local challenges and aspirations.

Investing in skills development empowers communities to become active participants in their own development. Education and training programs focusing on water management, renewable energy technologies and other relevant skills foster self-sufficiency and create a skilled workforce to drive future growth.

It needs no underscoring that local knowledge and practices are not simply relics; they can be a springboard for innovation. By combining this wisdom with cutting-edge technologies and scientific research, culturally appropriate and effective solutions can be developed. This requires that local communities, governments, NGOs and the private sector must work together, leveraging their unique resources, expertise and networks. This fosters trust, ensures a holistic vision for development and maximizes the impact of all efforts. Therefore, empowering local communities in arid regions is not just about charity, it's about unlocking their vast potential for resilience, innovation and stewardship. Their deep-rooted knowledge, combined with the right support and collaborative partnerships, can pave the way for a future where arid lands not only survive, but thrive.

Way Forward

The development of arid areas demands a multifaceted approach that addresses various aspects of sustainability, community empowerment and economic growth. Water management is critical, and strategies such as rainwater harvesting, desalination (considering costs and brine disposal), smart irrigation, and wastewater treatment are essential.

Empowering local communities through training and shared decision-making, while integrating indigenous knowledge, is vital for sustainable development. Sustainable agriculture, responsible tourism and promoting local crafts can create economic opportunities and improve livelihoods.

Renewable energy sources, such as solar and wind power, can reduce dependence on fossil fuels and contribute to a greener future. Combating desertification and preserving biodiversity are also crucial for the long-term environmental health of arid regions.

To implement these strategies effectively, a collaborative, three-tiered development approach is proposed. Village Management Committees for a cluster of 4–5 villages can empower communities to manage land use and project execution, while State Boards can advise and link local efforts. A National Authority can provide national guidance and international connections, ensuring a cohesive and coordinated effort.

State Agricultural Universities (SAUs) can be key partners in selecting drought-resistant local and native plants ideal for the environment. Local medicinal plants can offer additional benefits by greening the landscape and generating income for local communities through sales or value-added products. To ensure success, accessible nurseries with affordable, high-quality plants should be developed, along with training for locals in plant care and assigned responsibilities for planting, aftercare, and security. Water management and protection from extreme weather are crucial for sapling survival.

Knowledge dissemination through technology adoption, customized standard operating procedures (SOPs) and training programs involving SAUs, agriculture departments and forest departments can empower long-term sustainability. Public-private partnerships and international linkages can further strengthen these initiatives.

Combating desertification, preserving biodiversity and tackling malnutrition through nutritional parks can add to the project's holistic approach. Linking the project with tourism and wellness initiatives, such as yoga and meditation, can further incentivize participation, creating a thriving future for arid regions.

Shree Ann—Opportunity in Gujarat

A.R. Pathak[1]

The Food and Agriculture Organisation of the United Nations has declared year 2023 as the International Year of Millets as per the proposal of our Honourable PM Shri. Narendra Modiji, with following objectives.

- Raise awareness of the contribution of millets in food and nutritional security.
- Inspire stakeholders on improving sustainable production and quality of millets.
- Enhanced investment in research and development and extension services to achieve other two aims.

During 2018, Ministry of Agriculture and Farmers welfare recognized millets as "Nutri-cereals".

Of the nine millet crops, Pearl millet (Bajara), Sorghum and Finger millets (Ragi) are major whereas six millets viz; Kodo millet (kodara), Proso millet (cheno), Foxtail millets (Kang), Little millet (Vari), Brown yard millet (Samo/Moraiyo) and Browntop millets (lily kangli) are positive millets possessing antioxidant and minerals. Amarantus (Rajagra) is also a millet used during fast have antioxidants and steroids. The area under these crops was 35 million hectare which has reduced to only 15.48 million hectare during 2017–18. The total production of millets in our country is 50.9 million tons with productivity of 1239 kg/ha. Due to availability of irrigation, the area under wheat, paddy, cotton and sugarcane increased Inspite of these, India is a major producer with 41% contribution in world production hence there is great potentiality for export of these crops.

Millet crops are drought tolerance require less irrigation can be grown in light soil and climate resilient can withstand against abnormal weather conditions. Being a C4 plants more carbon absorption and sequestration hence improve soil health. From nutritional point of view, contains fibre, protein, vitamins minerals, antioxidants and Phyto chemicals higher than other cereals.

[1] Former Vice Chancellor, Junagadh Agricultural University and Navsari Agricultural University, Gujarat, India.

With respect to area and export of these crops, India ranked number one and second respectively. Honourable PM with foresight proposed UNO to declare IYM so our farmer gets benefit and increase their income.

These crops are climate resilient grown under rainfed condition with no or less chemical fertilizer and pesticides are best suited for sustainable agriculture. At present the market of millets is Rs. 742.4 crore which is expected to increase to the tune of Rs. 989.8 crores during 2025. Hence there is great opportunity for export and possibility of increasing income of our farmers. Therefore, there is now urgent need to take appropriate action for popularization and improvement of millets.

In Gujarat, pearl millet (Bjara), Sorghum (jowar) and Ragi (nagli) are mainly cultivated whereas other millets viz; kodara (kodo millet), kang (foxtail millet), vari (little millets), cheno (proso millet) and Rajagra (Amarantus) are also cultivated in more or less area but now their area under cultivation is also very meager. In comparison to 2013–14, the area under pearl millet and Sorghum has reduced to half. Similarly, cultivated area under Ragi and other millets also reduced to more than half. Hence Gujarat contribution in millet production in the country is only 7%. The area under these crops in Gujarat is 5.1 lakh hectare and production of 12 lakh tons with productivity of 2350 kg/ha – which is above national productivity.

Increasing Farmers Income through "Shree Ann"

The cultivated area of these crops in Gujarat is mainly in tribal parts in hilly and light soils. Valsad, Dang, Panchmahal, Tapi, Surat, Banaskantha and Sabarkantha growing minor millets and Sorghum. Whereas Pearl millet is grown in Mehsana, Ahmedabad, Surendranagar, Kheda, Anand, Junagadh, Jamnagar, Bhavnagar and Somnath district: There is a great need to increase production of these crops and enhance income our farmers. As International Year of Millets declared as well as health consciousness increased, there is great potentiality for export and domestic use which can increase farmers income. Therefore, action oriented programme for these is highly required. Some of the Points are as under:

- In Gujarat, awareness of Natural Farming is increasing and area is also increasing, there is great opportunity to grow millets under Natural Farming, as these crops require less fertilizer and tolerant to pest and diseases with less irrigation. Due to Natural Farmers will get better Produce available to consumers without chemicals.

- Easy availability of quality seeds of improved high yielding varieties of these crops to farmers.
- Evolving high yielding improved varieties and remunerative Package of practices through research involving conventional and modern technologies.
- Providing subsidy to seed producers and farmers for purchase of certified seeds. Though it is provided but not for local varieties which are better in taste and more preferred and demand by consumers, eg. Babarkot variety of Bajara; Nizer Gott, BP 53 and MalDandi in Sorghun, Desi red Colour in Ragi. For these highly preferred varieties a subsidy should be provided as a special case.
- Aware farmers about research recommendations immediately through FLDs, training etc. For this Farmers Training Centre and KVKs should organize programmes. Farmers should be informed about-crop production technologies and processing techniques as all crops require processing.

As these crops are climate resillient can be grown in all season. Therefore different zone wise area can be increased for suitable crop. The area expansion can be possible to grow these crops as intercrop or relay crop in main crop with early maturing varieties. More research in this direction is required.

- Most of millet crops require processing, cleaning, separating seeds etc. after harvesting, needs specific machines and training. NABARD and other banks should provide loan for purchase of such machinery.
- For different millet crops, area specific "Millet Village" can be develop. Department of Agril., KVKS can do this. One district-one crop scheme of GOI.
- GOI has announced many schemes on IYM. Farmers should be educated and make aware of these schemes for farmers benefit.
- FPOS for these crops in different area be established.
- MSP is for Bajara, Sorghum and Ragi, which should also be announced for other millet crops.
- Millets should be included in mid-day-meal scheme and PDS which will increase demand and help in malnutrition problem Too.
- The erception of millets as a poor man food, needs to be changed through marketing problem and capacity building.

Other orphan crops grown in tribal area which are good for nutritional and medicinal point of view e.g. Cassava, Yam, Niger, Giloy, Aloevera etc. for which research and development attention is required.

Agriculture and Carbon Sequestration: A Climate Change Mitigation Strategy

Odemari Mbuya[1] **and Kirit N. Shelat**[2]

Introduction

For better understanding of the chapter, knowing of terminologies below is essential.

- Weather refers to atmospheric conditions (temperature, precipitation, humidity wind speed, coldness, visibility) at a particular time in a specific location.
- Climate is the average of weather patterns on a specific area or region over a long period of time, usually 30 years or more.
- Greenhous gases (GHG) are gases that trap heat in the atmosphere, thereby causing global warming and climate change. Such gases include carbon dioxide (CO_2), methane (CH_4) and nitrous oxide (N_2O).
- Global warming is an increase of the Earth's average surface temperature due increased concentrations of GHG in the atmosphere.
- Climate change refers to the long-term changes of the Earth's climate.
- Climate variability refers to all variations in the climate that lasts longer than individual weather events.
- Mitigation is any action taken by governments, businesses, and people to reduce, sequester, or prevent GHG emissions.
- Adaptation refers to actions that help reduce vulnerability to the current or expected impacts of climate change.
- Resilience is the capacity of a community or environment to anticipate and manage dangerous climatic effect, recover, and transform after the ensuing shock, with minimal damage to societal wellbeing, economic activities, and the environment.
- Carbon sink is anything that absorbs more carbon from the atmosphere than it releases. Forests, wetlands, oceans, and soil are the world's largest carbon sinks.

---

[1] Center for Water Resources, Florida Agricultural and Mechanical University (USA).

[2] National Council for Climate Change, Sustainable Development and Public Leadership, India.

- Carbon sequestration is the capture and secure storage of carbon dioxide (CO_2) that would otherwise be emitted to or remain in the atmosphere.
- Agriculture or farming is the science, art, and practice of cultivating the soil to produce crops and raising livestock.
- Industrial revolution is the rapid development of industry that occurred in Britain, continental Europe, and the United States in the late 18th and 19th centuries, brought about by the introduction of machinery, growth of factories, and mass production of manufactured goods.

Green Revolution, Greenhouse Gas Emissions and Climate Change

Due to the Industrial Revolution, human activities have led to excessive use of fossil fuels, changes in land use and land cover patterns, which have inadvertently resulted into a sharp rise in concentration of greenhouse gases (GHG), such as CO_2, in the atmosphere (Figure 1). The concentration of CO_2 in the atmosphere has increased from 280 $\mu L\ L^{-1}$ (ppm) in 1860 to about 410 $\mu L\ L^{-1}$ at present, increasing at a rate of 0.45% per year, and it is expected to increase to 550 $\mu L\ L^{-1}$ by 2030 (IPCC, 2013). Since the beginning of the industrial era (1750), human activities have raised atmospheric concentrations of CO_2 by about 50% whereas the earth's average temperature has increased by 1.1°C. An

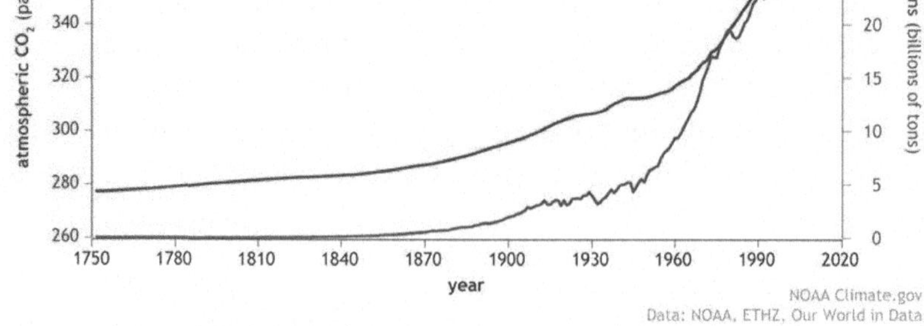

Figure 1: Concentrations of CO_2 in the Atmosphere and Annual Emissions (1750–2019)

increase in temperature beyond 1.5°C would cause many irreversible catastrophic environmental changes that pose serious threat to life on earth and human civilization.

There is ample evidence that increase of CO_2 concentration and other GHGs in the atmosphere accelerate global warming and triggers melting of glaciers and rise of sea level, which in turn triggers a chain reaction of frequent extreme weather events such as floods, typhoons, drought, and heat waves, seriously affecting the sustainability of ecosystem structure, functions, and services. The U.S. Energy Information Administration estimates that in 2019, the United States emitted 5,130 million metric tons of energy-related CO_2, while the global emissions of energy-related CO_2 totaled 33,621.5 million metric tons (USGS, 2019). There are two major ways to stop the increase of CO_2 concentration in the atmosphere, i) stop or reduce adding it to the atmosphere and/or ii) increase the ability of the earth to remove it from the atmosphere. Companies can use renewable energy sources to power factories and transport their products using fuel efficient trucks, cargo planes and ships to reduce the amount of CO_2 added to the atmosphere. On the other hand, we can increase the Earth's carbon sink by conserving forests, grasslands, peatlands, and wetlands where carbon is stored in plants and soils, thus protecting existing carbon sinks. Farming (agriculture) involves planting of crops (for food and fiber) and trees through production agriculture and forestry. Farming practices such as use of cover crops and crop rotation maintain soil health and make it effective carbon sink.

Carbon Sequestration and Climate Change

Reducing the amount of CO_2 in the atmosphere through carbon sequestration is the most effective and economical means of mitigating GHG effect on climate change. Plants are known to absorb sunlight for photosynthesis, converting water and CO_2 into carbohydrates. The Earth's carbon sink has been reduced through deforestation, change of land use and land cover, whereas carbon sources have been increased by excessive use of fossil fuels caused the Green Revolution (Figure 2).

The imbalance between carbon source and carbon sink created by human activities is the major cause of climate change. Increasing the carbon sink through vegetation is one way of offsetting the current source: sink imbalance. For example, reforestation of an area the size of the United States could reduce atmospheric CO_2 concentration by 25 percent, to the

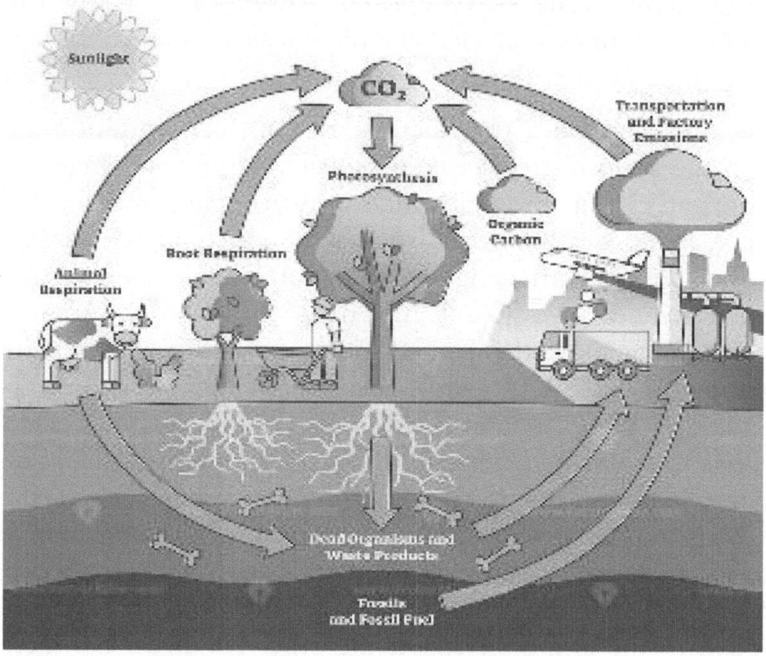

Figure 2: Carbon Cycle

lowest level in a century (Bastin *et al.*, 2019). Trees and agricultural crops are one option, but there are many other useful biological solutions. Algae are by far more effective than trees at removing CO_2 from the atmosphere (sequestration) and can be used to generate carbon-negative fuels, plastics, textiles, food, fertilizers, and more. Microalgae can absorb CO_2, nitrogen (N), phosphorus (P), and other components, while reducing GHG emissions. Microalgae can also play other roles such as purifying municipal wastewater and producing energy (biofuel).

Agriculture, Greenhouse Gases and Carbon Sequestration

Agriculture serves as both a source and sink for GHGs. The primary sources of GHGs in agriculture are the production of nitrogen-based fertilizers (NBFs), combustion of fossil fuels (coal, petrol, diesel, natural gas), waste management, and livestock enteric fermentation which releases methane (CH_4). In 2005, agriculture accounted for 10 to 12% of total global human caused emissions of GHGs (IPCC, 2007). On the other hand, agriculture (plants) removes CO_2 from the atmosphere through the processes of photosynthesis, converting water and CO_2 into organic carbon (biomass). In the United States, GHG emissions from agriculture account for 8% of all emissions and have increased since 1990 (Congressional Research Service, 2008). Conservation-tillage, cover-cropping, crop rotation and a suite of other best management practices (BMPs) can drastically increase the amount of carbon stored (sink) in soils.

In order to effectively reduce the current amount of CO_2 in the atmosphere, we must increase the carbon sink area and increase the carbon sequestration efficiency. Agriculture could be used to increase the carbon sink to mitigate climate change and restore soil health. Agricultural carbon sink could be increased by cultivating efficient carbon sequestration crops and trees, and by increasing the carbon sink area. Carbon sink area could be increased through the use of uncultivated and wasteland (e.g., desert, swamp, saline-alkali land, savanna, and wetland), and use of aquatic and soil microalgae.

Industrial Hemp and Algae as Model Crops for Carbon Sequestration

Although several crops could be used for carbon sequestration discussion, industrial hemp (*Cannabis sativa*) and soil microalgae (*Chlorella* spp and *Scenedesmus* spp) will be used as examples of high-efficiency carbon sequestration model plants. Viewed as an eco-friendly and highly sustainable crop, industrial hemp provides many environmental benefits. Hemp has exhibited superior GHG abatement over similar crops (140% and 540% greater than canola and sugar beet, respectively) in a similar field (Finnan and Styles, 2013). If hemp stalk is made into building materials, the lime binder could sequester CO_2 during hardening through carbonation. The final material that results is carbon neutral or even negative, sequestering from 6.67 to 136.65 kg CO_2 eq. m^{-3} (Barbieri, 2020).

Hemp is known for its carbon sequestration efficiency which is essential during this era of climate change. Hemp is one of the fastest sources of CO_2-to-biomass converter and can capture more CO_2 per hectare than other commercial crops, grasslands, or forests. Hemp is also listed as a natural way to clean up soil pollution. This phytoremediation method can harmlessly extract toxins and pollutants from soil and groundwater.

Usually industrial hemp grows to 3–5 m tall. It has an obvious carbon sink function and has a positive role in protecting the biological carbon cycle and improving the soil environment. During the growth cycle of more than 100 days, hemp can capture more than 25 tons of carbon dioxide per hectare, and part of the carbon dioxide can also be stored in the soil during the growth process. Due to the differences in climate environment and farmland management level, the effective number of hemp plants per square meter is between 20–30, the average dry weight is about 5.23 kg, the biological yield per hectare is about 15 tons, and the moisture content is 40%. The carbon density of hemp plants is about 0.45, and the carbon sequestration of hemp crop per hectare is about 6.75 tons, which is about 27.75 tons of carbon dioxide absorbed. The biological yield per unit area of cotton is about 7661.55 kg ha^{-1}, the carbon sequestration per hectare is about 3.83 tons, and the equivalent carbon dioxide absorption is about 14.04 tons. It can be seen that planting hemp has obvious carbon sequestration benefits than planting cotton (Adesina *et al.*, 2020).

In terms of ecological benefits, fiber hemp has a strong tolerance to heavy metals, and even when the cadmium content in the soil reaches 800 mg/kg, it has no obvious phytotoxic effect on the growth of fiber hemp. Metal ions selectively accumulate in the roots of plants, and only a small part is transported to stems and leaves. The single-layer adsorption capacity of hemp for chromium ion, copper ion, silver ion and cadmium ion in solution is 367 mg/kg, 1157 mg/kg, 89 mg/kg and 140 mg/kg, respectively. Therefore, hemp plants are excellent natural metal adsorbents, which have an effective cleaning effect on water bodies and soils (Nadia *et al.*, 2019). As early as 1998, ecological restoration scientist Slavik Dushenkov concluded that industrial hemp is a very effective regeneration plant; then, the land near the famous pollution site of Chernobyl nuclear power plant began to grow industrial hemp, and the land restoration effect is very significant.

It is well known that microalgae have a super high carbon sequestration capacity. Microalgae have the ability to fix carbon dioxide 10–50 times more than other terrestrial plants (Batista *et al.*, 2015). The ability of microalgae to absorb carbon dioxide is much higher than that of trees, and most of the oxygen in the atmosphere is provided by microalgae. Microalgae possess a well- known CO_2 concentration mechanism (CCM) for efficient photosynthesis by acquiring inorganic carbon even from very low atmospheric CO_2 concentrations (Whitton, 2012). Microalgae do not require arable land and can survive well in places where other crop plants cannot, such as saline water, land and wastewater (Searchinger *et al.*, 2008; Wang *et al.*, 2008). In addition, microalgae can also absorb exhaust gases such as CO_2 and NOx, SOx in fuel gas, inorganic and organic carbon, N, P, and other pollutants in agricultural, industrial and sewage wastewater sources (Chisti, 2007; Hu *et al.*, 2008; Pires *et al.*, 2012; Singh and Thakur, 2015). The simple cell structure and rapid growth of microalgae allow them to fix CO_2 10–50 times more efficiently than terrestrial plants (Li *et al.*, 2008; Khan *et al.*, 2009).

Compared to traditional crops, algae can produce nearly one hundred times more oil per acre, with a potential of producing 10,000 gallons/acre/year (Greenwell *et al.*, 2010). More important, algae strains could grow in uncultivated land (desert, swamp, saline-alkali land, savanna, wetland). When it comes to the environmental benefits, algae are the most potent microbe in CO_2 utilization and biological carbon fixation than other microbes and thus, help in mitigating the greenhouse effect. About 1 kg of micro-sized algae fixes 1.84 kg of atmospheric CO_2 (Gayathri *et al.*, 2021). Previous studies have proved that algae can efficiently absorb CO_2 through the activity of the "LCIB" protein in a low-concentration CO_2 environment, but scientists did not know what the specific environment and conditions were. According to the "Nihon Keizai Shimbun", a research team led by Associate Professor Takashi Yamano of Kyoto University found that the "LCIB" protein, which is closely related to the absorption of CO_2 can function in different parts of the chloroplast according to the concentration of CO_2 in the water to efficiently absorb CO_2 (Yamano *et al.*, 2014).

Hypergiant Industries, an artificial intelligence company in the United States, has developed an algal bioreactor, the EOS Bioreactor, that uses algae to absorb carbon dioxide from the atmosphere. The company claims algae is "one of nature's most efficient machines," and developers hope to pair it with AI systems to enhance its capabilities. The research team says its capacity is equivalent to planting 400 trees and is estimated to handle about two tons of carbon dioxide. With enough EOS devices, they can make entire cities net-zero carbon, or even negative, much faster than trees. That's the dream: clean air and livable cities for everyone right now.

Microalgae photosynthesis consumes a large amount of carbon sources such as carbon dioxide, and organic acids such as acetic acid, succinic acid, and citric acid after microbial decomposition can be used as carbon sources for microalgae to synthesize organic matter. Generally, the pH of microalgae is getting higher in the later stage, which is just to resist the process of soil acidification. It is reported that the changes of environmental pH, the amount of dust and the content of organic matter were similar to different algae species and under the same culture conditions. The effects of algal species change on pH and microbial increases were greater than those caused by substrate changes and light changes.

In order to effectively reduce the CO_2 content in the atmosphere, we must increase the carbon sink area and increase the carbon sequestration efficiency. Based on this, we propose to increase the carbon sink area of uncultivated land (desert, swamp, saline-alkali land, savanna, and wetland) through soil microalgae, and by planting fiber hemp, a crop with high carbon sequestration efficiency, to improve the carbon sink capacity in barren land where is unsuitable for other crops. At the same time, microalgae can improve the physical and chemical properties and organic matter content of the soil. Industrial hemp can also improve the soil by adsorbing heavy metals in the soil, achieving the effect of killing two birds with one stone.

Industrial hemp as a Climate Smart Commodity Crop

In addition to the removal of CO_2 from the atmosphere, industrial hemp has the potential to provide raw material for the emerging markets of new construction material (e.g., hempcrete and hemp plywood), bioplastics, biochar, livestock bedding, oil, and livestock feed, and thus may serve as a new lucrative source of income for farmers and other stakeholders.

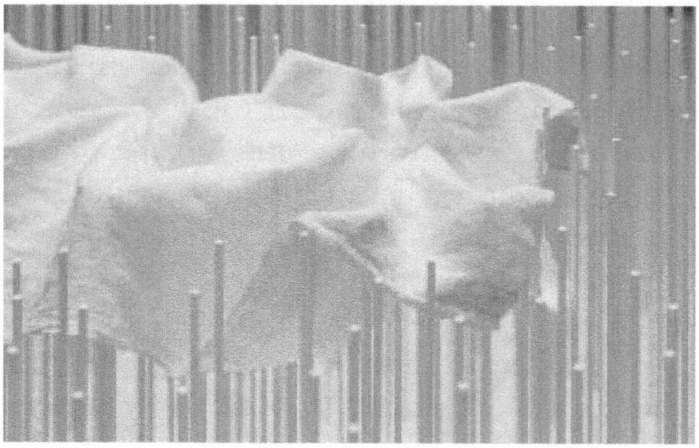

Figure 3: Hemp Fibre

The following products could be produced from industrial hemp as climate smart commodities; i) hemp fiber could be used for production of bioplastics, paper, cloth (Figure 3) and twine, ii) hemp biomass for manufacture of construction material (e.g., and hempcrete and hemp plywood) (Figure 4 and 5), providing markets for hemp farmers, and

iii) new foods on the menu for America and other countries (e.g., hemp burger and hemp crumbles) (Figure 6).

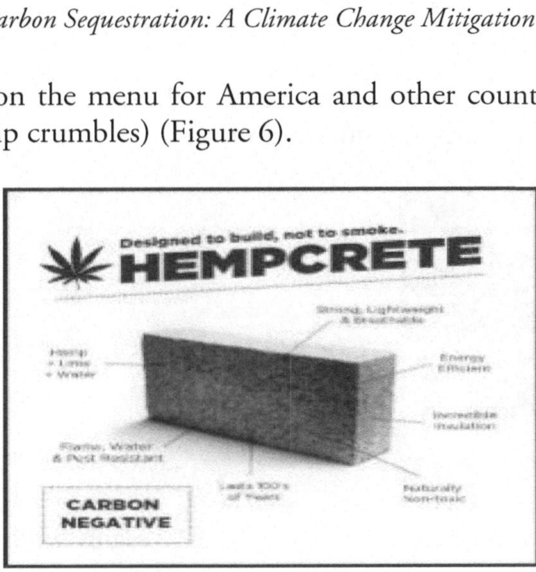

Figure 4: Hemp Biomass

Industrial hemp has many potential economic benefits, especially when we see the decreasing net farm income in the United States (US). Hemp has an estimated revenue of \$2,632 ha^{-1}, more profitable than traditional crops, like kenaf, switchgrass, and sorghum (Parvez *et al.*, 2021). In fact, the global market for hemp has reached the size of approximately \$ 6 billion, and was anticipated to double by 2020, largely due to the growth of US market (MPI, 2019). In the meanwhile, the US market for hemp-based products has a highly dedicated and growing demand base. Since 2011, the US total retail value of hemp products has increased from 10% to more than 20% annually. The US retail value is estimated at least \$688 million, a 20% growth over the 2015 retail value estimate of \$573 million (Thayer *et al.*, 2017).

Industrial hemp is extremely adaptable and can be grown in a variety of climatic conditions, and the requirements for soil conditions are not very strict. Whether it is a tropical area or an area above 3,000 meters above sea level, it can be said that as long as there are crops grown, industrial hemp can be cultivated. Hemp has a high carbon sequestration capacity (in terms of biomass) in areas with poor soil. It can improve the carbon sequestration capacity of plants in poor soil areas. At the same time, hemp can greatly improve the soil quality of heavy metal contaminated areas. Microalgae have high carbon sequestration capacity in decertified areas, and can improve soil physicochemical properties by increasing soil organic matter content, increasing soil pH and algal crust, especially in desert areas and acidified soils. Microalgae can improve the carbon sink capacity by greening the desert, increase the area of green space, and expand the scale of carbon sequestration.

Figure 4: Hempcrete

Figure 5: Plywood

Industrial Hemp as a Crop for Underserved Farmers

Industrial hemp (*Cannabis sativa*) can be grown by small and/or underserved farmers in many parts of the world, as an efficient carbon sequestration crop and climate-smart commodity crop. Products from a climate smart commodity crop like industrial hemp include, i) fiber for production of bioplastics, paper, cloth, and twine, ii) hemp biomass for manufacture of construction material like hempcrete and hemp plywood iii) livestock bedding and iv) new foods on the menu (e.g., hemp salad, hemp burger and hemp crumbles). These products provide new markets for farmers and other stakeholders.

A consortium of soil microalgae (*Chlorella* spp and *Scenedesmus* spp) can be added into the soil as soil amendment during planting. In addition to carbon sequestration, soil microalgae have the potential to improve soil dynamic properties of marginal and wasteland where underserved poor farmers live.

References

[1] Adesina, I.; Bhowmik, A.; Sharma, H. and Shahbazi, A. A Review on the Current State of Knowledge of Growing Conditions, Agronomic Soil Health Practices and Utilities of Hemp in the United States. Agriculture 2020, 10, 129. https://doi.org/10.3390/agriculture10040129

[2] Barbieri, Virginia (2020). Dissertation: Cereal By-products A Renewable and Sustainable Resource for Biobased Building Materials.

[3] Bastin, J.F.; Finegold, Y.; Garcia, C.; Mollicone, D.; Rezende, M.; Routh, D.; Constantin M. Zohner, C.M.; Thomas W. and Crowther, T.W.

(2019). The global tree restoration potential. Science, 365 (6448), DOI: 10.1126/science.aax0848.

[4] Chisti, Y. (2007). Biodiesel from microalgae. Biotechnol. Adv. 25, 294–306. doi: 10.1016/j.biotechadv.2007.02.001

[5] Finnan, John and David Styles. "Hemp: a more sustainable annual energy crop for climate and energy policy." Energy Policy 58 (2013): 152–162.

[6] Hu, Q., Sommerfeld, M., Jarvis, E., Ghirardi, M., Posewitz, M., Seibert, M., *et al.* (2008). Microalgal triacylglycerols as feedstocks for biofuel production: perspectives and advances. *Plant J.* 54, 621–639. doi: 10.1111/j.1365-313X.2008.03492.x

[7] Khan, S.A., Rashmi, Hussain, M.Z., Prasad, S. and Banerjee, U.C. (2009). Prospects of biodiesel production from microalgae in India. *Renew. Sust. Energy Rev.*, 13, 2361–2372. doi: 10.1016/j.rser.2009.04.005

[8] Li, Y., Horsman, M., Wu, N., Lan, C.Q. and Dubois-Calero, N. (2008). Biofuels from microalgae. *Biotechnol. Prog.*, 24, 815–820. doi: 10.1021/bp070371k

[9] Luo, Y.Q., SuB, Currie, W S, *et al.* Progressive nitrogen limitation of ecosystem responses to rising atmospheric carbon dioxide [J]. BioScience, 2004, 54(8): 731–739.

[10] MPI, 2019. Global hemp markets contents: Product and consumer landscapes. <https://www.mpi.govt.nz/dmsdocument/35142/direct>, (Accessed 25 February 2021).

[11] Pires, J.C.M., Alvim-Ferraz, M.C.M., Martins, F.G. and Simoes, M. (2012). Carbon dioxide capture from flue gases using microalgae: engineering aspects and biorefinery concept. *Renew. Sustain. Energy Rev.* 16, 3043–3053. doi: 10.1016/j.rser.2012.02.055

[12] Searchinger, T., Heimlich, R., Houghton, R.A., Dong, F., Elobeid, A., Fabiosa, J., *et al.* (2008). Use of US croplands for biofuels increases greenhouse gases through emissions from land use change. *Science* 319, 1238–1240. doi: 10.1126/science.1151861

[13] Singh, J. and Thakur, I.S. (2015). Evaluation of cyanobacterial endolith Leptolyngbya sp. ISTCY101, for integrated wastewater treatment and biodiesel production: a toxicological perspective. *Algal Res.*, 11, 294–303. doi: 10.1016/j.algal.2015.07.010

[14] Thayer, C., Burley, M., County, E. and Held, D. (2017). Industrial Hemp from Seed to Market. Cornell University, Harvest New York (Accessed 5 March 2021). <http://erie.cce. cornell.edu/resources/industrial-hemp-from-seed-to-market>.

[15] Wang, B., Li, Y., Wu, N. and Lan, C.Q. (2008). CO_2 bio-mitigation using microalgae. *Appl. Microbiol. Biotechnol*, 79, 707–718. doi: 10.1007/s00253-008-1518-y

[16] Whitton, B.A. (2012). Ecology of Cyanobacteria II: Their Diversity in Space and Time. Dordrecht: Springer. doi: 10.1007/978-94-007-3855-3.

[17] Yamano, T., Asada, A,, Sato, E. and Fukuzawa, H., Isolation and characterization of mutants defective in the localization of LCIB, an essential factor for the carbon-concentrating mechanism in Chlamydomonas reinhardtii. *Photosynth Res.,* 2014 Sep; 121(2—3):193–200. doi: 10.1007/s11120-013-9963-6. Epub 2014 Jan 3. PMID: 24384670.

[18] Zak, D.R., Pregitzer, K.S. and King, J.S. (2000). Elevated atmospheric CO_2 fine roots and the response of soil microorganisms: a review and hypothesis [J]. *New Phytologist,* 2000, 147(1): 201–222.

[19] Ziolkowska, J.R. and Simon, L. (2014). Recent developments and prospects for algae-based fuels in the US. *Renewable and Sustainable Energy Reviews,* 29, 847–853.

Future Agriculture at the Interface of Climate Change and Chemical Ecological and Ecosystems Level Perspectives of Bioresources Management: A Public Policy Outlook

R. Gopichandran[1]

One of the central features of public policy is the imperative of science at its core due to evidences it provides. This extends seamlessly into technology applications with special reference to benefits for the marginalized in particular. With respect to the former it is essential to explore domains that fine grain our understanding of resilience pathways, across the soil, water, air media as carriers of chemicals that mediate interactions and directly so on bioresources. The IPCC has periodically highlighted facets of climate resilient development pathways. Our understanding of planetary boundaries too has evolved; only to reinforce the call for rapid, yet sound mitigation and adaptation measures. Climate resilient agriculture is a case in point. It is essential to base related policies on principles of ecosystem level resilience, that are in turn driven by stochastic and non-stochastic processes and thresholds. System specific insights are therefore critical to assess individual, synergistic and antagonistic influences, duly respecting the time – scales over which they manifest. The present narrative is therefore a call to take note of emerging facets of ecology, especially of allelochemical and allelopathic pathways that could determine forms and functions of sustainable crop systems. The other aspect is about the impact of high levels of UV B exposure of crops due to depleting ozone layer.

We draw attention to recent developments in these three strands of ecology. The fourth strand is tools of contours of satellite imagery, drones and artificial intelligence for farmer engagement. These build on growing body of evidences about impacts of climate change on agriculture as

[1] Professor, NTPC School of Business, Noida, U.P. India.

emergencies and the call for resilience that cannot be overemphasized (FAO, 2024[1]). This calls also for innovative approaches to reduce vulnerabilities of ecosystems, productivity, livelihoods, agriculture – centered economy and ecosystem services that impact nutrition and food security. The final strand is about costs of mal adaptation and the urgent need to prevent backsliding on this font.

The first strand is about ecosystem level considerations. The Climate Smart Agriculture Sourcebook (FAO, 2024[2]) highlights the importance AFOLU and LULUCF for robust emission accounting and related methodologies to assist policy and plans for agriculture centered integrated mitigation and adaptation outcomes. The UNEP-DHI Centre[3] this year elaborated on the Water-Energy-Food-Ecosystems nexus with a scope to interpret tradeoffs. Adamos *et al.*[4] (2024) draw our attention to the EU policy alignments in this regard with implications for circular – economy. Dibyendu *et al.*[5] (2023) elaborated about nitrogen management to tackle climate change related challenges in agro-ecosystem productivity with implications for nitrogen – mediated allelochemical interactions as highlighted also by Moorman *et al.*[6] (2022). This feeds into the UN Global Campaign on Sustainable Nitrogen Management[7] and twins with phosphorus management (McDowell *et al.* 2024[8]) within the larger context of 'plant - interkingdom' allelopathic interactions (Schandry & Becker 2020[9]).

Anju & Gude [10](2022) and Gao *et al.*[11] (2023) reinforce the centrality of food energy and water for sustainable development and the need for holistic investigations on the nexus, notwithstanding complexities for a single – model integration. These aspects are true also of allelopathic interactions that influence the occurrence and distribution of feral hosts for bio - resources in ecosystems (Rao *et al.*[12] (2018); Bhowmik[13] (2022); Nagargade *et al*[14] (2024) and across media (Felpeto *et al.*[15] 2019). These in turn set the context for assessments of invasions of species from alternative habitats (Bakacsy *et al.*[16] 2024) & Lenda *et al.*[17] (2023).

The second strand emphasizes the reality that the above stated insights are indeed quite remarkable, albeit only as superstructures of insights. They should however be founded on a deeper appreciation of the chemical ecological dynamics of bioresources, that are at the core of agriculture productivity. Christiana *et al.*[18] (2022) emphasize the fact that climate change-associated stressors dynamics at the ecosystem level and across media are not adequately understood; specially to assess predictive capacities. This is especially so when such fine layering of impacts can be predominantly across at least seven levels, including tri-trophic interactions with implications for host preferences and avoidance, habitat selection and establishment of populations. These in turn appear to be influence eco-system services, driven by changes at the molecular level. Moisture films in soils could best exemplify microcosms within which carbonate equilibrium can be influenced by derivatives from photosynthesis and respiration, in turn could be determined by the gaseous environment.

Marcia *et al.*[19] (2023) present evidences of changes in leaf chemicals on attributable to warming and resultant herbivory. Guyer *et al.*[20] (2021) demonstrate changes due to climate variations in multi-trophic interactions in maize systems. Plant – root and related microbial dynamics too need greater attention especially when necromass carbon dominates. Cai *et al.*[21] (2023) & Liu *et al.*[22] (2024) argue about warming-induced changes in the quantum of available soil phosphorus, attributable to reduced microbial biomass. This could have implications the activity of soil extracellular enzymes and so the fate of necromass and soil organic carbon. These recent insights should be viewed against the backdrop of an excellent overview presented by Chitra *et al.*[23] (2021). Insights are presented by several others too about climate change related drought and heat stress, nitrogen metabolism, limitations and tolerance pertaining to salinity and water logging stresses, mineral nutrition and moisture stress and with reference to cotton, wheat, sorghum and other crop systems.

Other important areas are pollination ecology and volatiles that influence foraging that needs greater attention (Colazza & Rodriguez – Saona[24] 2023), much as Mutalipassi *et al.*[25] highlight the influence of novel compounds. Deep dives into the chemical ecological basis of productivity are therefore central to the development and implementation of preventive management strategies. Bobea *et al.*[26] (2021) present an econometric model to interpret weather effects "on global agricultural total factor productivity". The WEF[27] presents a framework of green returns much as it attracts attention to the susceptibility of five crops in particular to climate vagaries (WEF[28]).

With special reference to the Indian scenario (ICAR[29] and the PIB[30]) it is interesting to note that the OECD[31] in its 2023 – review elaborates on the economy – agriculture productivity nexus in the context of climate change. Shruti Mohapatra *et al.*[32] (2022) focus on the vulnerability of agribusiness using "exposure, sensitivity, and adaptive indicators" to establish the negative spiral caused by climate change on crop productivity, calling for a multipronged "social, technological, and economic" synergy to enhance adaptive capacity of farming systems including related stakeholders.

The Food and Land Use Coalition[33] presents a case from India about large-scale sustainable and regenerative agricultural practices dovetailed with sourcing and trade of commodities along with its Future Fit Food and Agriculture framework. Pratap Birthal *et al.*[34] too (2021) too establish negative impacts on crop yields, with implications for adjustments in land use, innovations in crop breeding for stress tolerance, integrated with appropriate management of land and water resources, as risk-mitigating mechanisms. An equally important review by Rama Rao *et al.*[35] (2022) focuses on major food crops using the district level climate projections. They call for a special focus on consequences of heat/temperature stress and related natural resource management programmes.

The third strand is about the convergence caused by ozone layer depletion and resulting high levels of UV B incident on crops. Very little is available in the public domain about the consequences of exposure of crops on their adaptive abilities, especially in the ozone – dip zone over the Indian sub-continent. One of the earliest snapshots on this aspect was presented by the USDA[36] (2003). Mmbando[37] (2023), Barnes *et al.*[38] (2023) and Bernado *et al.*[39] (2024) present some empirical evidences of defense mechanisms triggered by exposure that in turn influence yield.

The fourth strand is about the growing attention received by the interplay of satellite imagery – drones and AI as tools for citizen science based experiential learning and evidence-based communication. They are viewed as game-changers for the agriculture and allied sectors. This creates the scope for a systematic stock take of emerging trends world over that India can benefit from to appropriately augment her technology – enabled agriculture development pathways with; only to expedite outcomes, especially when SDGs are at the centre of India's vision for inclusive development.

Eleven important considerations in this context are India's own most recent analysis of opportunities[40] about 'digital transformation of farming", FAO's most recent "Adaptation, Biodiversity and Carbon Mapping Tool (ABC-Map)[41 - 43]" as part of the Agriculture, Food and Climate National Action Toolkit, the Food and Agriculture for Sustainable Transformation Partnership (FAST)[44] and about drone regulations[45] Observations from the The European Space Agency's World Cereal project[46] and IIASA's narrative on use of satellite data for food security at global scale; the global scale crop and irrigation monitoring[47] and cross cutting influences as in the CEOS Earth Observation Handbook[48] align with an important review by Hank *et al.* about space borne imaging spectroscopy in this context[49] NASA's ARSET[50] too presents its unique perspectives that complement the scope to build on Build on the impetus set by the recent global level assessment of food security and nutrition (FAO[51]). This is further substantiated by the Atoms4Food Initiative[52].

Accordingly it appears increasingly plausible that satellite imagery could rationalize agriculture and fisheries related options through nine tracks. They are: carbon sequestration: and thereby the choice of crops, other vegetation and seaweeds; preventive management of crop yield and health further implied by pest infestations and microbial infections and pest infestations; yield projections vis-à-vis age-correlated changes in related qualitative and quantitative profiles to guide harvests and related integrated resource management determining fertilization; locally adapted soil and water management practices that in turn determine fertilizer, herbicide and irrigation schedules; crop mapping and land use analysis to guide resource allocation, and market analysis; fish habitat mapping and monitoring including those of coral reefs, seagrass meadows, and mangroves. These in turn strengthen nature based management of spawning, feeding, and nursery grounds; fish stock estimates including size and distribution to set catch limits for sustainability; and ensure the sustainability of fisheries; detection of illegal fishing to strengthen enforcement and protection of fish stocks and monitor such ecosystems – related parameters as water/soil/habitat surface temperature, nutrients and currents – dynamics.

Drones in agriculture and allied sectors including livestock enhance efficiency, precision, and data-driven decision-making for farmers along five additional mutually reinforcing tracks. They are: crop monitoring with high-resolution cameras and multispectral sensors to capture detailed images of crops to assist timely preventive, ameliorative and augmentation practices; field mapping and analysis including data on soil conditions, elevation drainage patterns with implications for fertilizer and water application and related overall farm management practices; precision spraying and to reduce health externalities for crops and farmers; automated seed planting especially for cover crops and in challenging terrain and to improve planting efficiency, uniformity for better crop yield and monitor livestock herds movement health to improve grazing management, prevent disease outbreaks, and ensure overall well-being of animals.

Artificial Intelligence (AI) applications[53] in agriculture and allied sectors are gaining increasing attention. This is especially through three outlooks to enhance the scope for forecasts, vulnerability assessments and adaptation plans to reduce food security related divides[54] considering the FLAPP – Knowledge for Action[55] that call for innovations; interoperability, ease of data transfer, tests of models before market entry, forecast digital skills, design of sector-specific regulations and equity among stakeholders, reduce biases as deliberated by EU[56] and improvement of farming methods, reducing cost of inputs and services in reaching the unreached segments of farmers and to improve market access and facilitate integration of the smallholder farmer "into regional and global supply chains" as IFC's perspective[57]. A test bed should therefore be created as a microcosm for reality check building in the convergence posed by climate change impacts management presents. It should ideally present empirical evidences in the Indian context that can address all the facets stated about feasibility, limits, limitations and augments needed. This is aligned with FAO's call [58] for evidences, scenarios, design and appraisal of response options and to support multi stakeholder dialogues.

Upadhya and Arbuckle[59] highlight a basket of mal-adaptation actions in the USA, and with special reference to nutrient leakages and moisture losses through inappropriate practices in addition and attitudes to adaptation. The German Development Institute[60] presents a six-pronged approach to avoid mal-adaptation outcomes building on more than a dozen bio-physical and socio-economic indicators with a special focus on insurance. An interesting framework of parameters are derived through the Global Adaptation Mapping Initiative as stated by OECD[61] substantiating arguments abouts costs of inaction[62]. Hellin *et al.*[63] define a transformative adaptation outlook, closely aligned with the typology of adaptation costs defined by Kori[64], FAO[65] and Shipper[66] including the interplay of coping and no-response strategies. The essence of these prescriptions is the need for system – specific analysis of options to prevent back-sliding in particular.

References

[1] FAO 2024. FAO strategy on climate change 2022–2031. 50p. https://openknowledge.fao.org/server/api/core/bitstreams/f6270800-eec7-498f-9887-6d937c4f575a/content

[2] FAO 2013. Climate smart agriculture sourcebook 570p. https://www.fao.org/4/i3325e/i3325e.pdf

[3] UNEP DHI 2024. Policy note. Working with the Water-Energy-Food-Ecosystems (WEFE) nexus Summary and perspectives on the WEFE nexus for climate-resilient sustainable development. 04 p. https://unepdhi.org/wp-content/uploads/sites/2/2024/02/WEFE_nexus_policy_note_2024_UNEP-DHI_UNEP-CCC.pdf

[4] Adamos, G.; Serena, Caucci; Löic, Charpentier; Juliana, Chediek; Durk, Krol and Chrysi, Laspidou (2024). Exploring nexus policy insights for water-energy-food resilient communities. 14p. *Sustainability Nexus Forum* https://link.springer.com/article/10.1007/s00550-024-00534-0

[5] Dibyendu, C., Saurav, S., Tanumoy, B. and Debasish, S. (2023). Editorial: mitigation and adaptation strategies involving nitrogen management for enhancing ago-ecosystem productivity and resilience to future climate change 7: 1–3. https://www.frontiersin.org/articles/10.3389/fsufs.2023.1273821/full

[6] Moorman, J., Björn, H. and Tatjana, M. Hildebrandt (2022). Review News about amino acid metabolism in plant–microbe interactions 47 (10): 839 – 850. https://www.cell.com/action/showPdf?pii=S0968-0004%2822%2900169-4

[7] UN (2019). The Colombo Declaration https://www.inms.international/colombo-declaration/colombo-declaration

[8] McDowell, R.W., Pletnyakov, P. and HAygarth, P.M. (2024). Phosphorus applications adjusted to optimal crop yields can help sustain global phosphorus reserves. Nature Food 5: April 2024, 332–348. https://www.nature.com/articles/s43016-024-00952-9

[9] Schandray, N. and Becker, C. (2020). Allelopathic Plants: Models for Studying Plant–Interkingdom Interactions. *Trends in Plant Science,* 25(2): 176–185.
https://www.sciencedirect.com/science/article/pii/S1360138519303048

[10] Anju, V.N. and Gude, V.G. (2022). Chapter 16 – Food-energy-water nexus and assessment models. Water and Climate Change, Sustainable Development, Environmental and Policy Issues 2022, 317–330. https://www.sciencedirect.com/science/article/abs/pii/B9780323998758000185.

[11] Gao, D., Albert, S.C. and Fayyaz, A.M. (2024). A Systematic Review of Methods for Investigating Climate Change Impacts on Water-Energy-Food Nexus. Water Resources Management (2024). 38:1–43 https://link.springer.com/article/10.1007/s11269-023-03659-x

[12] Rao, A.N., Ravi, G.S., Gulshan, M. and Wani, S.P. (2018). Weed research issues, challenges, and opportunities in India. Crop Protection https://doi.org/10.1016/j.cropro.2018.02.003 from https://core.ac.uk/download/pdf/219475437.pdf.

[13] Bhowmik, P.C. (2022). Bioavailability of allelochemicals in soil environment under climate change: Challenges and perspectives. *Indian Journal of Weed Science* (2022) 54(4): 389–396. https://www.isws.org.in/IJWSn/File/2022_54_Issue-4_389-396.pdf

[14] Nagargade, M., Singh, M.K., Vishal, T., Prabhu, G., Anil, K.C., Kuldeep, R., Adarsh, K., Preeti, S. and Debalin, S. (2024). Ecological weed management and square planting influenced the weed management, and crop productivity in direct-seeded rice. *Scientific Reports. Nature Portfolio.* 14: 10356. 14p. https://www.nature.com/articles/s41598-024-56945-y

[15] Felpeto, A.B., Sylwia, S.W., Marek, K., Zofia, K. and Vitor, V. (2019). Temperature-dependent impacts of allelopathy on growth, pigment, and lipid content between a subpolar strain of Synechocystis sp. CCBA MA-01 and coexisting microalgae. *Hydrobiologia* (2019) 835:117–128 https://link.springer.com/article/10.1007/s10750-019-3933-8

[16] Bakaczy, L., Luca, V.K., Agnes, S., Krisztina, N.N., Andrea and Gabor, F. (2024). Investigation of the Allelopathic Effect of Two Invasive Plant Species in Rhizotron System. Life Basel 2024 Apr; 14(4): 475 https://www.ncbi.nlm.nih.gov/pmc/articles/PMC11051222/

[17] Lenda, M., Bastian, S., Piotr, S., Zuzanna, B.Z., Dawid, M., Renata, B., Franciszek, J., Agnieszka, B., Hugh, P.P. and Johannes, M.H.K. (2023). Multiple invasive species affect germination, growth, and photosynthesis of native weeds and crops in experiments. *Scientific Reports. Nature Portfolio,* 23:22146. https://www.nature.com/articles/s41598-023-48421-w

[18] Christiana, C.R., Mahasweta Solene, B., Paula, S., Patrick, F., Francois, V. and Jorg, D.H. (2022). Becoming nose-blind—Climate change impacts on chemical communication. Glob. Chang. Biol. 2022, Aug; 28(15): 4495–4505. https://www.ncbi.nlm.nih.gov/pmc/articles/PMC9321854

[19] Marcia, G.T., Valeria, P.O., Carolina, A., Alfredo, J.I. and Axel, M. (2023). Climate change related warming shifts in leaf chemical traits favor nutrition of the specialist herbivore *Battus polydamus archidamas. Frontiers in Ecology and Evolution,* 05 May 2023. 11p. https://www.frontiersin.org/articles/10.3389/fevo.2023.1152489/full

[20] Guyer, A.G., Cong van, D., Corina, M., Ricardo, ARM, Pierre M, Katja S, Lucie K, Günter Hoch, Ansgar K, Matthias E and Christelle AMR. 2021. Climate change modulates multitrophic interactions between maize, a root herbivore, and its enemies. Journal of Chemical Ecology (2021) 47:889–906 https://link.springer.com/article/10.1007/s10886-021-01303-9.

[21] Cai, M., Guang, Z., Bo, Z., Nan, C., Zhoutao, Juntao, Z., Xiaoqing, D., Yangjian and Yangjian, Z. (2023). Climate warming alters the relative importance of plant root and microbial community in regulating the accumulation of soil microbial necromass carbon in a Tibetan alpine

meadow. 2023 *Glob Chang Biol.* 29 (11) 3193–3204 https://onlinelibrary.wiley.com/doi/full/10.1111/gcb.16660

[22] Xiaofei, L., Ye, T., Jakob, H., Erika, S., Steve, K.K., Werner, B., Andreas, S. and Wolfgang, W. (2024). Long-term soil warming decreases soil microbial necromass carbon by adversely affecting its production and decomposition. Abstract as in https://onlinelibrary.wiley.com/doi/abs/10.1111/gcb.17379.

[23] Chitra, S., Padmavathi, C., Himabindu, P. and Naganna (2022). Molecules to ecosystems—recent trends in chemical ecology for combating biotic stresses in a changing climate, Chapter 13, 361–410, pp, In Climate Change and Crop Stress, Molecules to ecosystems, *Eds* Arun K. Shanker, Chitra Shanker, Anjali Anand & Maheshwari M. Elsevier Inc. Abstract as in https://www.sciencedirect.com/science/article/abs/pii/B9780128160916000110

[24] Colazza, S. and Cesar, R.S. (2023). Editorial: Insights in chemical ecology 2022. Frontiers in Ecology and Evolution 21 Feb. 2023. https://www.frontiersin.org/articles/10.3389/fevo.2023.1154019/full

[25] Mutalipassi, M., Riccio, G., Galasso, C., Zupo, V. and Greco, S. (2022). Editorial: Cyanobacterial and algal compounds: chemical ecology and biotechnological potentials Front. Mar.Sci. 24 Aug 2022, Sec. Biotechnology and Bioproducts Vol 9. https://www.frontiersin.org/articles/10.3389/fmars.2022.984160/full

[26] Bobea AO, Toby RA, Carlos MC, Robert GC and David BL, 2021 Anthropogenic climate change has slowed global agricultural productivity growth. Nature Climate Change. 11: 306 – 312. Abstract as in https://www.nature.com/articles/s41558-021-01000-1

[27] WEF 2023 Green Returns: Unleashing the Power of Finance for Sustainable Food Systems. White paper. 22p. https://www.weforum.org/publications/green-returns-unleashing-the-power-of-finance-for-sustainable-food-systems/

[28] WEF 2024 Climate Action – Extreme weather is driving food prices higher. These 5 crops are facing the biggest impacts https://www.weforum.org/agenda/2024/02/climate-change-food-prices-drought/ Feb 12 2024.

[29] ICAR Feb 2024 Indian farming. 10 Feb World Pulses Day. Nourishing Soils and People. 40p. https://icar.org.in/sites/default/files/2024-02/18%20Indian%20Farming%20February%202024.pdf

[30] PIB 05 March 2024 Ministry of Information and Broadcasting, 10 years of transformation: revolutionizing agriculture. https://pib.gov.in/PressNoteDetails.aspx?NoteId=151859&ModuleId=3

[31] OECD 2023 India. Chapter 14, In: Agricultural Policy Monitoring and Evaluation 2023: Adapting Agriculture to Climate Change, OECD Publishing, Paris, https://doi.org/10.1787/b14de474-en.

[32] Shruti M, Swati M, Heesup H, Antonio AM and Maria del CLM 2022, Climate change and vulnerability of agribusiness: Assessment of climate change impact on agricultural productivity https://www.ncbi.nlm.nih.gov/pmc/articles/PMC9645113/ Front Psychol. 2022; 13: 955622.

[33] The Food and Land Use Coalition. 2023 Food and land use context in India https://www.foodandlandusecoalition.org/country/india/

[34] Birthal SP, Jaweriah H, Negi DS and Bhan SC. 2021. Climate change and land-use in Indian agriculture. Land Use Policy, 109: October 2021, 105652 Abstract as in https://www.sciencedirect.com/science/article/abs/pii/S0264837721003756

[35] C.A. Rama Rao, B.M.K. Raju, Samuel Josily, A.V.M.S. Rao, R. Nagarjuna Kumar, M Srinivasa Rao, N Swapna, G Samba Siva, Y L Meghana, M Prabhakar and V K Singh 2022. Impact of climate change on productivity of food crops: a subnational level assessment for India. Environ. Res. Commun. 4 (2022) 095001. 17p. https://iopscience.iop.org/article/10.1088/2515-7620/ac8b68/pdf

[36] Gordon MH, Richard HG and Wei G 2003 Ultraviolet radiation and its impacts on agriculture and forests. 05p. https://www.fs.usda.gov/ne/newtown_square/publications/other_publishers/OCR/ne_2003heisler01.pdf

[37] Mmbando GS 2023 The recent relationship between ultraviolet-B radiation and biotic resistance in plants: a novel non-chemical strategy for managing biotic stresses. Plant Signal Behav. 2023; 18(1): 2191463 https://www.ncbi.nlm.nih.gov/pmc/articles/PMC10730183/

[38] Barnes, P.W., Robson, T.M., Zepp, R.G., Bornman, J.F., Jansen, M.A.K., Ossola, R., Wang, Q.-W., Robinson, S.A., Foereid, B., Klekociuk, A.R., Martinez-Abaigar, J., Hou, W.-C., Mackenzie, R. and Paul, N.D. (2023). Interactive efects of changes in UV radiation and climate on terrestrial ecosystems, biogeochemical cycles, and feedbacks to the climate system. Photochemical & Photobiological Sciences (2023) 22:1049–1091 https://link.springer.com/article/10.1007/s43630-023-00376-7

[39] Wallace de PB, Anne RS, Ellen MV, Saulo P, Laísa ZC, Guilherme AR deSouza, Deivisson P de Abreu, Letícia KOC, Felipe AA, Danilo FB, Fábio LP, Ricardo HBS, Miroslava R, José CR, Eliemar C and Weverton PR. 2024. UV-B reduction and excess: Management strategies regarding *Coffea* sp. Crop. Scientia Horticulturae 323: 1 Jan 2024, 112499. Abstract as in https://www.sciencedirect.com/science/article/pii/S0304423823006672

[40] Telecommunication Engineering Centre & ICAR 2024. Revolutionizing agriculture. The digital transformation of farming. TEC 31228:2024. A joint Technical Report. 147p, Released at the ITU/FAO Workshop on 18th March 2024, New Delhi. https://icar.org.in/sites/default/files/Circulars/TEC_ICAR_Revolutionizing%20Agriculture_The%20Digital%20Transformation%20of%20Farming%20%281%29%20%281%29.pdf

[41] FAO 2024 FAO launches "ABC-Map" geospatial app to boost country capacity to confront climate change https://www.fao.org/climate-change/news/news-detail/fao-launches--abc-map--geospatial-app-to-boost-country-capacity-to-confront-climate-change/en;

[42] FAO Geospatial information for sustainable food systems https://www.fao.org/geospatial/en/

[43] FAO Agroinformatics https://www.fao.org/agroinformatics/en

[44] FAO Food and Agriculture for Sustainable Transformation Partnership (FAST) https://www.fao.org/climate-change/action-areas/access-to-climate-finance/fast/en

[45] FAO 2024 Navigating Drone Regulations in Agriculture: Insights from Global Experts https://www.fao.org/e-agriculture/news/navigating-drone-regulations-agriculture-insights-global-experts

[46] ESA WorldCereal Towards a user-friendly and customizable global crop mapping system. WorldCereal Phase I to demonstrate feasibility of global crop mapping. https://esa-worldcereal.org/en,

[47] See, L., Gilliams, S., Conchedda, G., Degerickx, J., Van Tricht, K., Fritz, S., Lesiv, M., Bayas, J.C.L., Rosero, J., Tubiello, N.F. and Szantoi, Z. (2023). Dynamic global-scale crop and irrigation monitoring. *Nat Food* 4, 736–737 (2023). https://doi.org/10.1038/s43016-023-00841-7; as in https://www.nature.com/articles/s43016-023-00841-7

[48] Paganini, M., Petiteville, I., Ward, S., Dyke, G., Steventon, M., Harry, J. and Kerblat, F. (Eds.) 2018 CEOS Earth Observation Handbook, Satellite earth observations in support of SDGs. 114p. https://eohandbook.com/sdg/files/CEOS_EOHB_2018_SDG.pdf;

[49] Hank, T.B., Berger, K., Bach, H., Clevers, J.G.P.W., Gitelson, A., Tejada, P.Z. and Mauser, W. (2019). Spaceborne Imaging Spectroscopy for Sustainable Agriculture: Contributions and Challenges. Surveys in Geophysics (2019) 40:515–551 https://link.springer.com/content/pdf/10.1007/s10712-018-9492-0.pdf.

[50] NASA ARSET – Satellite remote sensing for agricultural applications. https://appliedsciences.nasa.gov/get-involved/training/english/arset-satellite-remote-sensing-agricultural-applications

[51] FAO, IFAD, UNICEF, WFP and WHO (2023). The State of Food Security and Nutrition in the World 2023. Urbanization, agrifood systems transformation and healthy diets across the rural-urban continuum. Rome, FAO. https://doi.org/10.4060/cc3017en FAO the State of Food Security and Nutrition in the World 316p. https://openknowledge.fao.org/server/api/core/bitstreams/8b27c570-2f8b-4350-8d5a-8e82432e6db7/content

[52] Nuclear News October 24, 2023. IAEA and FAO launch global food security initiative https://www.ans.org/news/article-5460/iaea-and-fao-launch-global-food-security-initiative/

[53] Mana, A.A., Allouhi, A., Hamrani, A., Rehman, S., Jamaoui, l el and. Jayachandran, K. (2024). Sustainable AI-based production agriculture: Exploring AI applications and implications in agricultural practices. Smart Agricultural Technology, Volume 7, March 2024, 100416 https://www.sciencedirect.com/science/article/pii/S2772375524000212.

[54] FAO. Artificial Intelligence best-practices in agriculture can help bridge the digital divide while tackling food insecurity https://www.fao.org/newsroom/detail/Artificial-Intelligence-best-practices-in-agriculture-can-help-bridge-the-digital-divide-while-tackling-food-insecurity/en

[55] FAO Agroinformatics https://www.fao.org/agroinformatics/training-and-resources/videos/flapp-(fao-food-loss-app)---knowledge-for-action/en

[56] European Parliament 2023. Artificial intelligence in the agri-food sector, Applications, risks and impacts. EPRS. STOA. 92p. https://www.europarl.europa.eu/RegData/etudes/STUD/2023/734711/EPRS_STU(2023)734711_EN.pdf

[57] Cook, P. and O'Neill, F. (2020). Artificial intelligence in agribusiness is growing in emerging markets. 8p. https://www.ifc.org/content/dam/ifc/doc/mgrt/emcompass-note-82-for-web.pdf

[58] FAO Water – energy – food nexus https://www.fao.org/land-water/water/watergovernance/waterfoodenergynexus/en/

[59] Upadhaya, S. and Arbuckle, J.G. (2021). Examining factors associated with farmers' climate – adaotive and maladaptive actions in the US Midwest. *Front. Clim.*, 28 June 2021 Sec. Climate risk management Vol. 3: 2021. 16p. https://www.frontiersin.org/journals/climate/articles/10.3389/fclim.2021.677548/full

[60] German Development Institute (2017). Risks of maladaptation: Climate insurance in agriculture. 4p. https://www.idos-research.de/uploads/media/BP_22.2017.pdf

[61] OECD (2023). Agricultural policy monitoring and evaluation 2023 Adapting agriculture to climate change 689p. https://www.oecd-ilibrary.org/sites/af245310-en/index.html?itemId=/content/component/af245310-en

[62] EEA (2023). Assessing the costs and benefits of climate change adaptation. 20p. https://www.eea.europa.eu/publications/assesing-the-costs-and-benefits-of

[63] Hellin, J., Fisher, E., Taylor, M., Bhasme, S. and Loboguerrero, A.M. (2023). Transformative adaptation: from climate-smart to climate-resilient agriculture. 8p. *CABI Agriculture and Bioscience* (2023) 4:30 https://cabiagbio.biomedcentral.com/articles/10.1186/s43170-023-00172-4

[64] Kori, S.D. (2023). A typology of climate adaptation costs for a smallholder maize farming system. *Climate Risk Management*, Vol. 40: 2023. 100517. https://www.sciencedirect.com/science/article/pii/S2212096323000438

[65] FAO Adapting to climate change https://www.fao.org/climate-smart-agriculture-sourcebook/concept/module-a2-adaptation-mitigation/chapter-a2-2/fr/

[66] ELF Shipper (2020). Maladaptation: When Adaptation to Climate Change Goes Very Wrong. One Earth 3: 409–414 https://www.cell.com/one-earth/pdf/S2590-3322(20)30483-8.pdf

Agroforestry for the Future: Driving Sustainability, Securing Livelihoods, and Combating Climate Change

Sanjay Deshmukh[1]

Introduction

Definition and Importance of Agroforestry

Agroforestry represents a transformative approach to agriculture, where trees and shrubs are integrated into crop and livestock systems. This integration is not merely a juxtaposition of different species but a synergistic interaction that leverages the unique benefits of each component. These deliberate arrangement and management elements result in a multifunctional landscape that enhances ecological, economic, and social sustainability. Agroforestry systems are designed to optimize the production of food, fiber, fuel, and fodder while maintaining ecological balance and promoting biodiversity.

The importance of agroforestry extends beyond its immediate benefits to farmers. It is critical in mitigating climate change through carbon sequestration, reducing soil erosion, and improving water retention and quality. By enhancing biodiversity, agroforestry systems contribute to the resilience of agricultural landscapes, making them more capable of withstanding and recovering from environmental stresses. This integrative approach also offers economic advantages by diversifying income sources for farmers and creating employment opportunities in rural areas.

Historical Context and Evolution

Integrating trees with agricultural activities is deeply rooted in human history. Traditional agroforestry systems have been practiced for centuries by Indigenous communities worldwide. In India, systems like home gardens, where diverse species of plants, including trees, crops, and medicinal herbs, are cultivated together, have been an integral part of

[1] Professor & Head, University Dept. of Life Sciences; formerly, Vice-Chancellor, University of Mumbai, Mumbai 400098, India. E-mail: sanjaydeshmukh@mu.ac.in

rural livelihoods. These traditional practices were based on a deep understanding of local ecosystems and the interdependencies between different species.

Over time, scientific research has expanded our understanding of agroforestry and its benefits. Early research focused on documenting traditional practices and understanding their ecological and economic impacts. This knowledge was then used to develop improved agroforestry models incorporating modern agricultural techniques and technologies. The evolution of agroforestry reflects a continuous process of learning and adaptation, where traditional knowledge is combined with scientific innovation to create more sustainable and productive agricultural systems.

In the modern era, agroforestry has gained recognition as a viable strategy for sustainable development. International organizations, governments, and research institutions have invested in agroforestry research and development, leading to the dissemination of best practices and the scaling up of successful models. In India, the National Agroforestry Policy, launched in 2014, aims to promote the adoption of agroforestry through supportive policies, research, and capacity-building initiatives.

Relevance to Sustainable Agriculture in India

India's agricultural sector is at a crossroads, facing significant challenges such as land degradation, water scarcity, and the impacts of climate change. Conventional agricultural practices, which often rely on monocultures and chemical inputs, have contributed to these problems. In contrast, agroforestry offers a holistic solution that addresses these challenges while enhancing agricultural productivity and sustainability.

Agroforestry systems improve soil health by reducing erosion, enhancing soil structure, and increasing nutrient cycling. Trees in agroforestry systems act as windbreaks, reducing wind erosion and protecting crops. Their roots stabilize the soil, reducing water erosion, and their leaf litter adds organic matter to the soil, improving its fertility and water-holding capacity. This is particularly important in India, where soil erosion and nutrient depletion are major concerns.

Water management is another critical benefit of agroforestry. Trees enhance groundwater recharge by increasing infiltration and reducing surface runoff. They also help maintain soil moisture, crucial for crop growth, especially in rainfed agricultural areas. By improving water retention and quality, agroforestry systems contribute to more resilient agricultural landscapes.

Agroforestry has a significant advantage in the mitigation of climate change. Trees sequester carbon, reducing the concentration of greenhouse gases in the atmosphere. Additionally, agroforestry systems create microclimates that can moderate temperature extremes and provide more favorable growing conditions for crops and livestock. This is particularly relevant in India, where climate variability significantly threatens agricultural productivity.

Economically, agroforestry diversifies farmers' income sources by providing timber, fruits, nuts, and fodder products. This diversification reduces the economic risks associated with crop failures and market fluctuations. Furthermore, establishing and managing agroforestry systems creates employment opportunities in rural areas, contributing to poverty alleviation and rural development.

Agroforestry promotes community engagement and empowerment socially. Local communities are involved in the planning and managing of agroforestry systems, fostering a sense of ownership and collective response-bility. Women and youth, in particular, play a crucial role in agroforestry initiatives, contributing to social inclusion and gender equity.

Agroforestry Systems and Practices

Types of Agroforestry Systems

Agri-Silviculture

Integrating Crops with Trees: Agri-silviculture is an agroforestry system in which trees and crops are grown together on the same land. This integration can be simultaneous or sequential, depending on the specific needs and management practices. The trees provide numerous benefits, including improved soil fertility through nitrogen fixation, enhanced microclimates that protect crops from extreme weather, and increased biodiversity. Common examples include intercropping leguminous trees with cereal crops, where the trees fix atmospheric nitrogen and improve soil fertility, thus benefiting the adjacent crops.

The success of agri-silviculture largely depends on carefully selecting tree species compatible with the crops. Fast-growing, nitrogen-fixing trees such as *Gliricidia sepium* and *Leucaena leucocephala* are often preferred due to their ability to improve soil fertility rapidly. Additionally, the spacing and arrangement of trees and crops must be managed to minimize competition for light, water, and nutrients.

Silvo-Pastoral Systems

Combining Forestry with Livestock Grazing: Silvo-pastoral systems integrate trees, forage, and livestock on the same land, providing a diversified farming system that enhances productivity and sustainability. This system offers improved forage quality, enhanced animal welfare, and better nutrient cycling. The trees in silvo-pastoral systems provide shade and shelter for livestock, reducing heat stress and increasing animal productivity. Moreover, tree leaves and pods can serve as additional fodder during lean periods.

Successful silvopastoral systems require the selection of tree species that are beneficial to livestock and resilient to grazing pressure. Trees like Prosopis juliflora and Albizia lebbeck are commonly used due to their drought tolerance and nutritional benefits to livestock. Management practices in silvopastoral systems focus on balancing tree density and ensuring adequate forage availability for livestock.

Agri-Silvo-Pastoral Systems

Integrating Crops, Livestock, and Trees: Silver systems represent a more complex form of agroforestry, combining crops, livestock, and trees in a single integrated system. This multifunctional approach maximizes land use efficiency and provides diverse products and services. The interactions between the different components can improve soil health, increase biodiversity, and enhance resilience to climatic and economic shocks.

The key to successful agri-silvo-pastoral systems is carefully managing the interactions between crops, livestock, and trees. This involves selecting species that complement each other and adopting management practices that optimize the benefits of each component. For instance, integrating nitrogen-fixing trees with forage crops can enhance soil fertility, while livestock can help control weeds and provide manure that enriches the soil.

Home Gardens - Multi-layered, Multi-species Systems

Home gardens are traditional agroforestry systems commonly found in tropical and subtropical regions. They are characterized by their multi-layered structure, with different species of trees, shrubs, and herbaceous plants arranged in vertical and horizontal layers. Home gardens are highly diverse, often including fruit trees, medicinal plants, vegetables, and ornamental species.

The multi-layered structure of home gardens mimics natural ecosystems, promoting high levels of biodiversity and ecological stability. These systems provide various products, including fruits, vegetables, herbs, and fuelwood, contributing to household food security and income. Additionally, home gardens play a crucial role in conserving traditional knowledge and plant genetic resources.

Species Selection and Management

Criteria for Selecting Tree Species

The selection of tree species in agroforestry systems is a critical decision influencing the system's overall productivity and sustainability. The criteria for selecting tree species include ecological compatibility, economic value, and social acceptability. Trees must be compatible with the local climate, soil, and cropping systems. They should also provide valuable products such as timber, fruits, fodder, or medicinal compounds.

Nitrogen-fixing trees, such as *Acacia, Leucaena, and Sesbania* species, are often preferred for their ability to improve soil fertility. Fast-growing species such as Eucalyptus and Populus, which can provide quick returns, are popular choices. However, the selection must consider potential negative impacts, such as allelopathy, where certain tree species may inhibit the growth of adjacent crops.

Management Practices for Optimal Growth and Yield

Effective management practices are essential to realizing the full benefits of agroforestry systems. These practices include proper tree spacing and arrangement, timely pruning and thinning, and effective pest and disease management. Proper spacing ensures that trees do not compete excessively with crops for light, water, and nutrients. Pruning and thinning help maintain the desired tree structure and promote the growth of trees and crops.

Soil fertility management is another crucial aspect, which may involve using organic amendments such as compost and green manure. Integrated pest management (IPM) strategies can be employed to control pests and diseases, minimizing the reliance on chemical pesticides. Regular monitoring and adaptive management are necessary to address any emerging issues and optimize the system's performance.

Case Studies of Successful Species Integration

Several case studies from various regions in India illustrate the successful integration of tree species in agroforestry systems. These examples demonstrate the diverse benefits of agroforestry, including improved soil fertility, enhanced crop yields, and additional resources for farmers.

(a) **Southern India—*Melia dubia* (Malabar Neem) with Maize and Sorghum:** In southern India, the integration of *Melia dubia* (Malabar Neem) with crops such as maize and sorghum has shown remarkable success. The fast-growing Melia dubia provides high-quality timber and significantly improves soil fertility, enhancing crop yields. Farmers have reported a substantial increase in maize and sorghum production, all thanks to the improved soil structure and nutrient cycling facilitated by *Melia dubia*. This system offers economic benefits through timber sales and is a testament to the potential of sustainable agricultural practices.

(b) **Northeast India—Alder (*Alnus nepalensis*) with Traditional Crops:** In the hilly regions of Northeast India, the integration of Alder (*Alnus nepalensis*) with traditional crops like rice and millet has been a beacon of success. Alder trees, with their nitrogen-fixing ability, not only enrich the soil but also play a crucial role in supporting the growth of adjacent crops. This agroforestry system has significantly increased crop yields, reduced the need for chemical fertilizers, and improved soil conservation, substantially impacting the environment and local farmers' livelihood.

(c) **Western Ghats—Areca Nut (*Areca catechu*) with Black Pepper (*Piper nigrum*)**: In the Western Ghats, farmers have successfully integrated Areca Nut (*Areca catechu*) plantations with Black Pepper (*Piper nigrum*). Areca nut trees provide a support structure for the climbing black pepper vines. This intercropping system maximizes land use efficiency and increases farm income by producing two valuable crops on the same land. The deep root systems of areca nut trees also help in soil stabilization and water retention, enhancing overall farm sustainability.

(d) **Central India—Teak (*Tectona grandis*) with Soybean and Wheat:** Integrating Teak (*Tectona grandis*) with soybean and wheat crops has proven beneficial in Central India. Teak trees, known for their high-quality timber, are planted along the borders of fields. These trees provide an additional income source from timber sales and offer shade and wind protection for the crops, leading to improved growth conditions. Farmers have observed better soil moisture retention and reduced soil erosion, contributing to higher soybean and wheat yields.

(e) **Northern India—Poplar (*Populus deltoides*) with Sugarcane:** In Northern India, particularly in the states of Uttar Pradesh and Punjab, integrating Poplar (*Populus deltoides*) with sugarcane has been a successful agroforestry practice. Poplar trees are planted in rows within sugarcane fields. The fast-growing poplars provide timber and pulpwood, while the intercropped sugarcane benefits from the improved microclimate the trees create. This system has led to enhanced sugarcane yields, additional income from timber, and better land utilization.

These case studies from India showcase the successful integration of various tree species with agricultural crops, demonstrating the multifaceted benefits of agroforestry. Farmers can enhance productivity, improve soil health, and contribute to sustainable agricultural development by adopting and promoting such practices.

Agroforestry Models

Traditional Models

Indigenous Practices and Local Knowledge: Traditional agroforestry models in India are deeply rooted in indigenous practices and local knowledge, developed over generations to adapt to specific environmental and socio-economic conditions. These models offer sustainable solutions that emphasize biodiversity, ecological balance, and the integration of multiple species. Here are five examples of traditional agroforestry systems in India:

(a) **Taungya System in Northeast India:** The 'Taungya' system, originally from Southeast Asia, has been adopted in the northeastern states of India, such as Assam and Nagaland. Farmers grow agricultural crops in this system between rows of young forest trees. This practice not only provides food and income during the early years of tree plantation but also helps manage weeds and maintain soil fertility. As the trees mature, they provide timber and other forest products, ensuring long-term benefits.

(b) **Zabo System in Nagaland:** The 'Zabo' system, practiced by the 'Chakhesang' tribe in Nagaland, integrates livestock rearing, forestry, and agriculture. This system involves the construction of water harvesting structures on hilltops to collect rainwater, which is then used to irrigate terraced fields below. Trees are planted around the terraces to prevent soil erosion and provide fodder, fuel, and timber. This holistic approach ensures sustainable land use, water conservation, and diversified livelihoods.

(c) **Agro-Horticulture in Himachal Pradesh:** In the mountainous regions of Himachal Pradesh, traditional agro-horticulture systems combine the cultivation of fruit trees with annual crops and livestock. Apple orchards are often intercropped with vegetables, pulses, and medicinal plants. The integration of fruit trees provides a steady income, improves soil fertility through leaf litter, and reduces soil erosion on steep slopes. This system also supports biodiversity and enhances the resilience of farming communities.

(d) **Home Gardens in Kerala:** Home gardens, or 'Kudumbashree', in Kerala represent a traditional agroforestry model where multi-layered, multi-species gardens are cultivated around homesteads. These gardens typically include fruit trees, vegetables, spices, medicinal plants, and fodder species. This diverse system ensures food security, income generation, and ecological balance year-round. Home gardens also promote the conservation of indigenous plant varieties and traditional knowledge.

(e) **Parkland Agroforestry in Rajasthan:** In the arid regions of Rajasthan, the traditional parkland agroforestry system involves the scattered planting of trees such as *Prosopis cineraria* (Khejri), *Acacia nilotica* (Babool), and *Azadirachta indica* (Neem) in agricultural fields. These trees provide multiple benefits, including shade, fodder, fuelwood, and soil fertility enhancement through nitrogen fixation and organic matter addition. The presence of trees also helps protect crops from wind erosion and provides a habitat for wildlife.

These traditional agroforestry models in India highlight indigenous practices' wisdom and relevance in contemporary sustainable agriculture. Preserving and promoting these systems can enhance ecological resilience, improve livelihoods, and contribute to sustainable development.

Modern Models

Innovative Approaches and Technological Integration: Modern agroforestry models in India incorporate innovative approaches and technological advancements to enhance productivity and sustainability. These models leverage improved tree species, precision agriculture techniques, and advanced management practices to optimize agroforestry systems. Here are five examples of modern agroforestry models in India:

(a) **Clonal Forestry in Tamil Nadu:** In Tamil Nadu, techniques are employed to produce high-yielding and disease-resistant varieties of tree species such as Eucalyptus and Casuarina. These clonal plantations are integrated with crops like groundnut and pulses. Using genetically improved clones ensures uniform growth, higher biomass production, and increased economic returns for farmers. Clonal forestry also contributes to carbon sequestration and soil conservation, enhancing sustainability.

(b) **Agroforestry with GIS and Remote Sensing in Andhra Pradesh:** In Andhra Pradesh, integrating Geographic Information Systems (GIS) and remote sensing technology has revolutionized agroforestry planning and monitoring. These tools help optimize the spatial arrangement of trees and crops, monitor soil health, and assess crop performance. Farmers use GIS-based maps to make informed decisions about tree species selection, planting density, and water management. This technological approach improves productivity and resource efficiency in agroforestry systems.

(c) **Precision Agroforestry in Punjab:** Precision agriculture techniques are being applied to agroforestry systems in Punjab to enhance productivity and sustainability. Farmers use advanced sensors, drones, and automated irrigation systems to monitor and manage their fields. For instance, soil moisture sensors and drip irrigation systems ensure optimal water use for trees and crops. This technology integration reduces water consumption, improves crop yields, and enhances the overall efficiency of agroforestry practices.

(d) **Climate-Smart Agroforestry in Maharashtra:** In Maharashtra, climate-smart agriculture frameworks are being incorporated into agroforestry models to enhance resilience against climate change. Farmers integrate drought-tolerant tree species such as teak and bamboo with crops like millet and legumes. These climate-resilient agroforestry systems improve soil health, increase water retention, and provide diversified income sources. Adopting climate-smart practices ensures the sustainability of agricultural landscapes in the face of changing climatic conditions.

(e) **Integrated Tree-Crop-Livestock Systems in Karnataka:** In Karnataka, innovative agroforestry models integrate trees, crops, and livestock into a cohesive system. Farmers plant nitrogen-fixing tree species such as Gliricidia and Leucaena in rows within crop fields and use the pruned biomass as fodder for livestock. This system enhances soil fertility and crop yields and provides a sustainable feed source for livestock. The livestock integration adds an additional revenue stream and improves farming systems' sustainability and resilience.

These examples of modern agroforestry models in India demonstrate how innovative approaches and technological integration can enhance agricultural landscapes' productivity, sustainability, and resilience. By adopting these modern practices, farmers can achieve higher economic returns, improve resource efficiency, and contribute to sustainable agricultural development.

Mixed Models

Combining Traditional Wisdom with Modern Techniques: Mixed agroforestry models in India blend traditional knowledge with modern innovations to create sustainable and productive systems. These models harness the ecological benefits and resilience of traditional practices while incorporating the efficiency and productivity of contemporary techniques. Here are five examples of successful mixed agroforestry models in India:

(a) ***Home Gardens in Kerala—Traditional Practices with Modern Horticultural Techniques:*** In Kerala, traditional home gardens, known for their biodiversity and multi-layered structure, are being enhanced with modern horticultural techniques. These home gardens have become more productive and diverse by incorporating improved irrigation systems, high-yielding vegetable varieties, and organic fertilizers. Farmers benefit from increased food security, additional income from surplus produce, and enhanced ecological stability.

(b) ***Silvopastoral Systems in Rajasthan—Traditional Practices with Modern Forage Management:*** In Rajasthan, traditional silvopastoral systems, which integrate trees with livestock grazing, are being improved with modern forage crop management techniques. Farmers plant fast-growing, drought-resistant fodder trees like *Prosopis cineraria* (Khejri) alongside modern forage crops. This integration enhances livestock productivity, improves soil health through better nutrient cycling, and provides a sustainable source of fodder even in arid conditions.

(c) ***Agroforestry in Himachal Pradesh—Traditional Apple Orchards with Modern Soil Fertility Management:*** In Himachal Pradesh, traditional apple orchards are being combined with modern soil fertility management practices to enhance productivity. Farmers use organic mulching, composting, and biofertilizers to improve soil health and apple yields. Integrating drip irrigation systems also ensures efficient water use, reducing water stress and increasing fruit quality. This mixed model supports sustainable apple production while preserving traditional orchard practices.

(d) ***Agroforestry in Karnataka—Indigenous Practices with Modern Crop Varieties:*** In Karnataka, Indigenous agroforestry practices, such as intercropping trees with staple crops, are being enhanced with modern crop varieties and techniques. Farmers plant traditional trees like Pongamia and Neem alongside improved millets, legumes, and vegetable varieties. Modern pest management and fertilization methods boost crop yields and income while maintaining the ecological benefits of traditional intercropping systems.

(e) ***Agroforestry in Uttarakhand—Traditional Mixed Cropping with Modern Agroforestry Techniques:*** In Uttarakhand, traditional mixed cropping systems, where farmers grow multiple crops together, are being integrated with modern agroforestry techniques. Farmers incorporate nitrogen-fixing trees like Alder and multipurpose species like Bamboo within their mixed crop fields. Modern practices such as contour planting and terracing prevent soil erosion and enhance water retention. This combination improves soil fertility, increases crop yields, and ensures sustainable land management.

These mixed agroforestry models in India illustrate how combining traditional wisdom and modern techniques can lead to sustainable and productive agricultural systems. By leveraging the strengths of both approaches, farmers can achieve higher productivity, improved resource efficiency, and greater ecological resilience.

Ecological and Environmental Benefits of Agroforestry

Agroforestry, the intentional integration of trees and shrubs into crop and livestock systems, offers many ecological and environmental benefits. It is crucial in promoting sustainable agriculture by enhancing biodiversity, improving soil health, aiding in water management, and contributing to climate change mitigation. This section delves into the key ecological and environmental benefits of agroforestry systems.

Soil Health Improvement

Soil Erosion Control

One of the primary benefits of agroforestry is its ability to control soil erosion. Trees and shrubs are natural barriers against wind and water erosion, stabilizing the soil with their root systems. The presence of trees reduces the velocity of surface runoff, thereby decreasing the erosive force of water. This is particularly beneficial in sloped or hilly areas, where soil erosion can be severe.

Agroforestry practices such as contour planting, where trees are planted along the contours of the land, significantly reduce soil erosion. Various regions have used this method to preserve topsoil, enhance soil fertility, and maintain agricultural productivity. Studies have shown that agroforestry systems can reduce soil loss by up to 90% compared to conventional monoculture systems.

Nutrient Cycling and Soil Fertility Enhancement

Agroforestry systems improve soil fertility through enhanced nutrient cycling. Trees and shrubs contribute organic matter to the soil through leaf litter, root turnover, and pruning residues. This organic matter decomposes, enriching the soil with essential nutrients such as nitrogen, phosphorus, and potassium.

Nitrogen-fixing trees, such as species from the genera *Acacia*, *Albizia*, and *Leucaena*, play a critical role in enhancing soil fertility. These trees have symbiotic relationships with nitrogen-fixing bacteria, which convert atmospheric nitrogen into a form that plants can use. This process not only improves the fertility of the soil but also reduces the need for chemical fertilizers.

Furthermore, trees in agroforestry systems' deep root systems can access nutrients from deeper soil layers unavailable to shallow-rooted crops. These nutrients are then brought to the surface through leaf fall and root exudates, making them accessible to crops and enhancing soil fertility.

Biodiversity Conservation

Habitat Creation and Wildlife Corridors

Agroforestry systems contribute to biodiversity conservation by creating habitats for various plant and animal species. The structural complexity of agroforestry systems, with multiple layers of vegetation, provides diverse habitats that support a wide range of organisms. This includes birds, insects, mammals, and soil microorganisms.

Tree cover in agricultural landscapes acts as wildlife corridors, connecting fragmented habitats and allowing species to move and disperse. This is particularly important in areas where natural forests have been cleared for agriculture, as agroforestry systems can serve as refuges for wildlife. The preservation of biodiversity within agricultural landscapes enhances ecosystem stability and resilience.

Enhancing Flora and Fauna Diversity

Introducing diverse tree and shrub species in agroforestry systems increases the overall biodiversity of agricultural landscapes. This diversity includes flora and fauna, as the varied vegetation provides different niches and resources for various organisms. The presence of trees can attract pollinators, beneficial insects, and other wildlife, which can enhance crop productivity and ecological balance.

Agroforestry systems often incorporate indigenous tree species, which are well adapted to local conditions and support native biodiversity. The conservation of these species is crucial for maintaining genetic diversity and ecosystem health. Additionally, agroforestry practices can help restore degraded lands, bringing back native vegetation and associated wildlife.

Climate Change Mitigation

Carbon Sequestration Potential

Agroforestry systems have significant potential for carbon sequestration, capturing and storing atmospheric carbon dioxide (CO_2). Trees and shrubs in agroforestry systems sequester carbon through photosynthesis, storing it in their biomass (trunks, branches, leaves) and the soil. This reduces the amount of CO_2 in the atmosphere, mitigating climate change.

Research indicates that agroforestry systems can sequester between 2 and 9 metric tons of CO_2 per hectare per year, depending on the tree species, management practices, and local conditions. This makes agroforestry a valuable strategy for meeting global climate targets. The integration of trees into agricultural landscapes reduces greenhouse gas emissions and enhances the resilience of farming systems to climate change impacts.

Microclimate Regulation

Trees and shrubs in agroforestry systems play a vital role in regulating microclimates. They provide shade, reduce wind speeds, and maintain soil moisture levels, creating a more favorable environment for crops and livestock. This microclimate regulation can mitigate the effects of extreme weather events such as heatwaves, droughts, and heavy rains, which are becoming more frequent due to climate change.

The shading effect of trees reduces soil and air temperatures, which can protect crops from heat stress and reduce evaporation rates. This is particularly important in regions with high temperatures and limited water availability. Additionally, trees' windbreak effect can prevent wind erosion and reduce crop damage from strong winds.

Water Management

Enhancing Groundwater Recharge

Agroforestry systems enhance groundwater recharge by improving soil structure and increasing water infiltration. Trees' deep root systems create channels in the soil, facilitating water movement into deeper layers. This process replenishes groundwater reserves essential for sustaining agriculture and rural livelihoods.

Tree cover also reduces surface runoff, allowing more water to percolate into the soil. Agroforestry practices such as alley cropping and contour planting are particularly effective in enhancing water infiltration and reducing runoff. By promoting groundwater recharge, agroforestry systems help ensure a reliable water supply for crops and livestock, especially during dry periods.

Reducing Surface Runoff and Soil Moisture Conservation

Trees and shrubs in agroforestry systems reduce surface runoff by intercepting rainfall and slowing water movement across the landscape. This reduces the risk of soil erosion and nutrient loss, which can degrade agricultural land. The improved soil structure and increased organic matter content in agroforestry systems enhance soil moisture retention, providing a more stable crop water supply.

Mulching, a common practice in agroforestry systems, further aids soil moisture conservation. Organic mulch from pruned tree branches and leaf litter covers the soil, reducing evaporation and maintaining moisture levels.

This is particularly beneficial in semi-arid and arid regions, where water scarcity is a major constraint to agricultural productivity.

Socio-Economic Impacts of Agroforestry

Agroforestry is an environmentally sustainable agricultural practice with immense potential for socio-economic benefits. This section explores the diverse socio-economic impacts of agroforestry, focusing on livelihood diversification, food security enhancement, community empowerment, and case studies illustrating these benefits.

Livelihood Diversification

Additional Income Sources for Farmers

Agroforestry systems provide farmers with multiple income streams, reducing their reliance on a single crop and enhancing economic resilience. Integrating trees and shrubs with crops and livestock allows for the production of various goods such as timber, fuelwood, fruits, nuts, fodder, and medicinal plants. These products can be sold in local and regional markets, generating additional farmer revenue.

For instance, cultivating fruit trees such as mango, guava, and citrus alongside staple crops provides a steady income from fruit sales. Similarly, timber and fuelwood from fast-growing tree species like Eucalyptus and Leucaena can be harvested and sold, providing farmers with a valuable source of income during off-seasons or periods of crop failure.

In many regions, agroforestry has proven to be a viable alternative to traditional farming, especially in marginal lands where conventional agriculture may not be economically feasible. Farmers can better withstand market fluctuations, climatic uncertainties, and other economic challenges by diversifying their income sources.

Employment Opportunities in Agroforestry Activities

The establishment and maintenance of agroforestry systems create employment opportunities in rural areas. Activities such as tree planting, pruning, harvesting, and processing require labor, providing jobs for local communities. This is particularly important in regions with high unemployment rates, where agroforestry can contribute to rural development and poverty alleviation.

Agroforestry also promotes the development of value-added industries, such as producing fruit preserves, essential oils, and timber products.

These industries can provide additional employment opportunities and stimulate local economies. Moreover, the skills and knowledge gained from working in agroforestry activities can enhance the employability of rural populations, empowering them with new capabilities and opportunities.

Food Security Enhancement

Contribution to Food Production and Nutritional Security

Agroforestry systems enhance food security by increasing the diversity and availability of food products. Combining trees and crops provides a more stable and reliable food source, as tree crops can be produced even during annual crop failure periods. This diversified production system reduces the risk of food shortages and enhances the resilience of farming systems to climatic and market shocks.

Incorporating fruit, nut, and vegetable trees in agroforestry systems improves nutritional security by providing a variety of vitamins, minerals, and other essential nutrients. For example, Moringa trees (*Moringa oleifera*), commonly used in agroforestry systems, are known for their highly nutritious leaves, pods, and seeds. Similarly, the inclusion of leguminous trees such as *Acacia* and *Albizia* can enhance protein intake through their edible seeds and leaves.

Resilience against Market and Climate Shocks

Agroforestry systems contribute to the resilience of farming communities by buffering against market and climate shocks. The diversification of crops and tree species reduces farmers' dependence on a single commodity, which can be vulnerable to price volatility and market fluctuations. By having multiple products to sell, farmers can mitigate the impact of price drops in one commodity by relying on the income from others.

In terms of climate resilience, agroforestry systems offer several advantages. Trees and shrubs can moderate microclimates, protecting crops from extreme weather conditions such as heatwaves, frosts, and heavy rains. Trees also provide shade and reduce soil temperature, enhancing crop growth and yield during hot and dry periods.

Furthermore, as discussed in previous sections, agroforestry systems improve soil health and water management, making agricultural lands more resilient to droughts and floods. This increased resilience is crucial for maintaining food production and ensuring food security in the face of climate change.

Community Empowerment

Engagement and Participation of Local Communities

Agroforestry initiatives often involve the active participation of local communities, fostering a sense of ownership and empowerment. Engaging communities in planning, implementing, and managing agroforestry systems can make these initiatives more sustainable and effective. Community involvement ensures the systems are tailored to local needs and conditions, enhancing their acceptance and success.

Participatory approaches, such as farmer field schools and community-based training programmes, are commonly used to promote agroforestry. These approaches empower farmers with the knowledge and skills to establish and manage agroforestry systems. They also facilitate the exchange of experiences and best practices among farmers, fostering a collaborative learning environment.

Role of Women and Youth in Agroforestry Initiatives

Agroforestry offers significant opportunities to empower women and youth in rural areas. Women, who often play a crucial role in household food production and natural resource management, can benefit from the income and nutritional security provided by agroforestry systems. Cultivating high-value crops such as fruits, nuts, and medicinal plants can enhance women's economic independence and social status.

Youth involvement in agroforestry initiatives can address the challenges of rural unemployment and migration. By providing employment opportunities and skills training, agroforestry can attract young people to agriculture and reduce the migration of rural youth to urban areas. This engagement is essential for the sustainability of agroforestry systems, as the next generation of farmers will be responsible for their continuation and expansion.

Case Studies

Success Stories from Various Regions in India

Numerous success stories from different regions in India illustrate the socio-economic benefits of agroforestry. These examples highlight how integrating trees with crops and livestock can lead to increased income, improved soil fertility, and enhanced resilience to environmental challenges.

(a) **Farm Forestry Project in Gujarat—Eucalyptus and Acacia Integration:** In Gujarat, the Farm Forestry Project has successfully encouraged farmers to integrate fast-growing tree species such as Eucalyptus and Acacia with their crops. This agroforestry model has increased income from timber sales, improved soil fertility, and enhanced resilience to droughts. The trees provide shade and windbreaks, reducing crop stress and increasing agricultural productivity. Additionally, deep-rooted trees help in maintaining soil moisture and preventing erosion.

(b) **Home Gardens in Kerala—Diverse Food Products and Biodiversity Preservation:** The traditional practice of home gardens, known as agroforests, has significantly succeeded in Kerala. These multi-layered gardens provide households diverse food products, fuelwood, and medicinal plants. Home gardens contribute to food and nutritional security, reduce household expenses, and preserve biodiversity by conserving indigenous plant species. This model has inspired similar initiatives in other parts of India, demonstrating the viability of integrating traditional practices with modern needs.

(c) **Wadi Model in Maharashtra—Horticulture and Forestry Integration:** The Wadi model, implemented in tribal areas of Maharashtra, integrates horticulture and forestry to improve livelihoods and environmental sustainability. Farmers cultivate fruit trees such as mango and cashew alongside forestry species like teak and bamboo. This system provides multiple income sources from fruit sales and timber, enhances soil fertility through organic matter from tree litter, and improves water retention in the soil. The Wadi model has significantly improved the socio-economic status of tribal communities.

(d) **Agroforestry in Odisha—Casuarina and Rice Integration:** Farmers in Odisha's coastal regions have integrated Casuarina trees with rice cultivation to combat soil salinity and improve crop yields. Casuarina trees act as windbreaks and reduce soil erosion, creating a favorable microclimate for rice growth. The leaf litter from the trees adds organic matter to the soil, enhancing its fertility. This agroforestry model has increased rice productivity, provided timber for construction, and improved the overall resilience of farming systems to coastal weather conditions.

(e) **Silvopastoral Systems in Rajasthan—Acacia and Forage Crop Integration:** In the arid regions of Rajasthan, silvopastoral systems have successfully integrated Acacia trees with forage crops like *Cenchrus ciliaris* (buffel grass) and *Stylosanthes hamata* (stylo). Acacia trees provide shade and reduce heat stress for livestock, while the forage crops offer nutritious feed. This integration improves soil fertility through nitrogen fixation by Acacia and organic matter from forage crops. The silvopastoral system has increased livestock productivity, provided additional income from timber, and improved soil health in the arid landscape.

These success stories from various regions in India demonstrate the diverse socio-economic benefits of agroforestry. By integrating trees with crops and livestock, farmers can achieve increased income, improved soil fertility, and enhanced resilience to environmental challenges, contributing to sustainable agricultural development.

Lessons Learned and Best Practices

The experiences from these and other agroforestry projects offer valuable lessons and best practices for future initiatives. Key factors contributing to the success of agroforestry systems include:

(a) ***Participatory Approaches:*** Involving local communities in the planning and managing of agroforestry systems ensures their acceptance and sustainability. Participatory approaches foster a sense of ownership and empower communities with the knowledge and skills needed for successful implementation.
(b) ***Species Selection:*** Choosing the right tree and crop species is crucial for the success of agroforestry systems. Species should be well adapted to local conditions, provide multiple benefits, and complement each other in terms of growth requirements and ecological functions.
(c) ***Integrated Management:*** Effective management practices, such as pruning, mulching, and intercropping, are essential for optimizing the productivity and sustainability of agroforestry systems. Integrated management approaches should consider the interactions between trees, crops, and livestock to maximize benefits and minimize competition.
(d) ***Market Access:*** Ensuring access to agroforestry product markets is critical for these systems' economic viability. This includes developing value chains, improving market infrastructure, and providing farmers with market information and support.

(e) ***Policy Support***: Supportive policies and institutional frameworks are necessary to promote agroforestry adoption and scaling up. This includes providing financial incentives, technical assistance, and research and development support for agroforestry initiatives.

Agroforestry systems offer a range of socio-economic benefits that contribute to sustainable rural development. By diversifying income sources, enhancing food security, empowering communities, and providing employment opportunities, agroforestry can improve the livelihoods of rural populations and build resilient agricultural systems. The success stories from various regions in India demonstrate the potential of agroforestry to transform agricultural landscapes and promote sustainable development. By learning from these experiences and implementing best practices, we can harness the full potential of agroforestry for the benefit of farmers, communities, and the environment.

Policy and Institutional Support for Agroforestry

The successful implementation and sustainability of agroforestry systems are significantly influenced by the policy and institutional framework within which they operate. This component delves into the critical aspects of policy and institutional support, examining national policies and frameworks, institutional mechanisms, funding and investment, and regulatory and legal aspects. Such support is crucial for mainstreaming agroforestry and ensuring its integration into national agricultural and environmental strategies.

National Policies and Frameworks

Overview of Agroforestry Policies in India

India has recognized the importance of agroforestry in addressing food security, environmental sustainability, and rural livelihoods. The National Agroforestry Policy 2014 was a landmark initiative to promote the practice nationwide. This policy provides a comprehensive framework for agroforestry, addressing various aspects such as research and development, extension services, and market access.

The policy encourages the integration of agroforestry into broader agricultural policies and programmes. It emphasizes the need for a coordinated approach involving multiple stakeholders, including government agencies, research institutions, non-governmental organizations, and the private sector. By providing a clear and supportive framework, the policy aims to create an enabling environment for farmers to adopt agroforestry practices.

Key Government Programmes and Schemes Supporting Agroforestry

Several government programmes and schemes have been launched to support adopting and scaling up agroforestry in India. Notable among these is the Sub-Mission on Agroforestry (SMAF) under the National Mission for Sustainable Agriculture (NMSA). SMAF focuses on promoting tree plantation on farmlands, providing financial assistance for establishing agroforestry systems and supporting capacity-building initiatives for farmers and extension workers.

Another significant programme is the 'Green India Mission', which aims to enhance forest and tree cover nationwide. This mission recognizes the role of agroforestry in achieving its objectives and supports agroforestry interventions, particularly in degraded and marginal lands.

The Mahatma Gandhi National Rural Employment Guarantee Act (MGNREGA) also offers opportunities to promote agroforestry. By integrating tree planting and maintenance activities into MGNREGA, the government has created a dual benefit of employment generation and environmental conservation.

Institutional Mechanisms

Role of Research Institutions and Universities

Research institutions and universities advance agroforestry through research, innovation, and capacity-building. Institutions such as the Indian Council of Agricultural Research (ICAR) and its affiliated institutes have been at the forefront of agroforestry research in India. These institutions research various aspects of agroforestry, including species selection, management practices, and the socio-economic impacts of agroforestry systems.

Universities offering agricultural and forestry education have incorporated agroforestry into their curricula, producing a new generation of scientists and practitioners with the knowledge and skills to advance the field. Collaborative research programmes involving international institutions have further enriched the knowledge base and facilitated the exchange of best practices.

Extension Services and Capacity Building Programmes

Extension services are critical for transferring knowledge and technologies from research institutions to farmers. The National Agroforestry Policy underscores the need for robust extension services to promote adopting agroforestry practices. Krishi Vigyan Kendras (KVKs), agricultural extension centers established by ICAR, play a significant role. KVKs conduct training programmes, demonstrations, and on-farm trials to educate farmers about the benefits and techniques of agroforestry.

Capacity-building programmes targeted at farmers, extension workers, and local communities are essential for successfully implementing agroforestry systems. These programmes provide training on various aspects of agroforestry, including species selection, nursery management, planting techniques, and post-harvest processing. By building local capacity, these programmes ensure the sustainability and scalability of agroforestry initiatives.

Funding and Investment

Public and Private Sector Investment in Agroforestry

Adequate funding and investment are crucial for developing and scaling up agroforestry systems. Both public and private sector investments play a significant role in this context. Public investment, primarily through government programmes and schemes, provides the initial impetus for farmers to adopt agroforestry. These investments are often in grants, subsidies, and technical assistance.

Private sector investment is equally important for the long-term sustainability of agroforestry. Companies involved in agro-based industries, such as food processing and timber production, are vested in promoting agroforestry. By investing in agroforestry projects, these companies can secure a sustainable supply of raw materials while contributing to environmental conservation and rural development.

Innovative financing mechanisms, such as public-private partnerships (PPPs), can leverage the strengths of both sectors. PPPs can mobilize additional resources, enhance the efficiency of project implementation, and create market linkages for agroforestry products.

Financial Incentives and Subsidies for Farmers

Financial incentives and subsidies are essential for encouraging farmers to adopt agroforestry practices. These incentives can offset the initial costs of establishing agroforestry systems and provide a buffer against potential risks. The National Agroforestry Policy advocates for providing financial incentives to farmers, particularly smallholders and marginal farmers.

Subsidies for inputs such as seeds, seedlings, fertilizers, and irrigation equipment can significantly reduce the financial burden on farmers. Credit facilities with favorable terms and conditions can also provide the necessary capital for farmers to invest in agroforestry. Insurance schemes that cover risks associated with tree crops, such as pests, diseases, and adverse weather conditions, can further encourage farmers to diversify their farming systems with trees.

Regulatory and Legal Aspects

Land Tenure and Ownership Issues

Land tenure and ownership issues are critical for successfully implementing agroforestry. Secure land tenure gives farmers the confidence to invest in long-term agroforestry systems. Conversely, insecure land tenure can discourage farmers from planting trees, as they may not be assured of reaping the benefits of their investments.

The National Agroforestry Policy highlights the importance of addressing land tenure issues to promote agroforestry. It calls for reforms to secure land rights for farmers, particularly those practicing agroforestry on marginal and communal lands. These reforms can include land titling, leasing arrangements, and recognizing customary land rights.

Legal Framework Governing Tree Planting and Harvesting

A supportive legal framework is essential for successfully implementing agroforestry. This includes regulations governing the planting, managing, and harvesting of trees on farmland. In many regions, restrictive laws and regulations have hindered agroforestry adoption by imposing cumbersome tree felling and transport procedures.

The National Agroforestry Policy advocates for the simplification of these regulations to facilitate the adoption of agroforestry. It calls for harmonizing forestry and agricultural policies to create a conducive environment for farmers to plant and manage trees. Additionally, the policy emphasizes the need for capacity building among regulatory agencies to ensure the effective implementation of these reforms.

Policy and institutional support are fundamental to the success and sustainability of agroforestry systems. A comprehensive and supportive policy framework and robust institutional mechanisms can create an enabling environment for farmers to adopt and benefit from agroforestry. Adequate funding, investment, financial incentives, and subsidies can provide the necessary resources to establish and scale up agroforestry systems. Finally, addressing land tenure and ownership issues and creating a supportive legal framework can remove barriers to adopting agroforestry.

By building on the progress made through initiatives such as the National Agroforestry Policy and various government programmes, India can harness agroforestry's full potential to achieve its food security, environmental sustainability, and rural development goals. The experiences and lessons learned from these initiatives can serve as valuable guidance for future efforts, ensuring that agroforestry becomes an integral part of the country's agricultural landscape.

Challenges and Future Directions

The road ahead is fraught with challenges and opportunities when pursuing sustainable agricultural development through integrated farming systems (IFS). This section explores the multifaceted landscape of implementation challenges, outlines critical research and development needs, and envisions future prospects and opportunities for scaling up agroforestry practices to meet national development goals.

Challenges in Implementation

Integrated farming systems, encompassing agroforestry among other components, face a spectrum of challenges that span technical, socio-economic, and environmental dimensions. Though formidable, these challenges present opportunities for innovation and collaboration across disciplines and sectors.

Technical and Knowledge Barriers

At the forefront of challenges lie technical and knowledge barriers that hinder the widespread adoption of integrated farming systems. Farmers and practitioners often lack access to comprehensive information and technical know-how to implement agroforestry practices effectively. Variability in agroecological conditions further complicates the application of standardized practices, necessitating context-specific approaches tailored to local realities.

Addressing technical barriers demands robust extension services, farmer-to-farmer knowledge-sharing networks, and capacity-building initiatives that empower stakeholders with practical skills in agroforestry management. Drawing upon interdisciplinary expertise, collaborative research endeavors are essential to generate contextually relevant innovations and adaptive management strategies that optimize the synergy between crops, trees, and livestock.

Socio-Economic Constraints

Socio-economic constraints pose significant challenges to adopting integrated farming systems, particularly in resource-constrained settings. Limited access to credit, land tenure insecurity, and inadequate market linkages undermine farmer confidence and investment in agroforestry initiatives. Moreover, the opportunity costs of transitioning from conventional to integrated farming systems deter widespread adoption among risk-averse farmers.

Supportive policy frameworks and institutional mechanisms are imperative to mitigate socio-economic constraints. Inclusive policies that incentivize sustainable land use practices provide access to microfinance, and foster market integration for agroforestry products can stimulate economic diversification and enhance livelihood resilience. Strengthening farmer cooperatives and producer organizations enables collective bargaining power, facilitating equitable distribution of benefits derived from integrated farming systems.

Environmental Limitations

Environmental limitations, including climate variability, soil degradation, and water scarcity, pose formidable challenges to the sustainability of integrated farming systems. Climate change exacerbates these challenges, altering agro-ecological dynamics and necessitating adaptive strategies that enhance ecosystem resilience and resource use efficiency.

Integrating climate-smart practices within agroforestry systems- such as drought-resistant tree species, water-efficient irrigation technologies, and soil conservation measures- is crucial to mitigate environmental risks and sustain agricultural productivity. Embracing agroecological principles promotes biodiversity conservation, carbon sequestration, and watershed management, reinforcing the ecological integrity of farming landscapes.

Research and Development Needs

Addressing the complex challenges facing integrated farming systems requires concerted research and development efforts to advance scientific knowledge, technological innovations, and policy support frameworks.

Priority Areas for Agroforestry Research

Critical research priorities in agroforestry encompass a broad range of topics essential for enhancing the effectiveness and resilience of integrated farming systems. These include understanding the intricate ecological interactions within agroforestry systems, optimizing the synergies between trees, crops, and livestock, and evaluating the socio-economic impacts of these practices on rural communities.

(a) ***Ecological Interactions:*** Elucidating the complex ecological interactions within agroforestry systems is fundamental to their optimization. Research should focus on the relationships between different plant species, including competition for resources like light, water, and nutrients and beneficial interactions such as nitrogen fixation and pest suppression. Understanding these dynamics can help design systems that maximize productivity and ecological benefits.

(b) ***Tree-Crop-Livestock Synergies:*** Optimizing the synergies between trees, crops, and livestock is another critical area of research. Studies should investigate how different species combinations can be managed to enhance overall system productivity and resilience. This includes exploring the potential of trees to provide shade and shelter for livestock, using crop residues as fodder, and the benefits of animal manure for soil fertility.

(c) ***Socio-Economic Impacts:*** Evaluating the socio-economic impacts of integrated farming practices is essential for understanding their broader implications for rural development. Research should assess how agroforestry systems contribute to income diversification, food security, and community empowerment. This involves analyzing the economic benefits of timber and non-timber forest products and the social and cultural values associated with traditional agroforestry practices.

(d) ***Integrated Pest and Disease Management:*** Developing integrated pest and disease management strategies tailored to agroforestry systems is crucial for maintaining plant health and productivity. Research should focus on identifying effective biological control agents, understanding pest and disease dynamics in mixed-species systems, and developing sustainable management practices that minimize the use of chemical inputs.

(e) ***Soil Fertility and Nutrient Cycling:*** Enhancing soil fertility through improved nutrient cycling is vital for the sustainability of agroforestry systems. Studies should explore the role of different tree species in nutrient cycling, the impact of organic matter from tree litter on soil health, and the potential for agroforestry practices to sequester carbon and improve soil structure.

(f) ***Resilient Tree Species:*** Developing and promoting resilient tree species that can thrive under diverse agroecological conditions is paramount. Research should focus on breeding and selecting tree varieties that are resistant to pests and diseases, tolerant of drought and other climatic stresses, and capable of providing multiple ecosystem services.

Role of Technology in Advancing Agroforestry

Harnessing digital technologies, remote sensing, and precision agriculture tools can revolutionize monitoring and management practices within agroforestry landscapes. These technologies offer new ways to gather and analyze data, improve decision-making, and enhance the overall efficiency of agroforestry systems.

(a) ***Digital Technologies:*** Digital technologies, including mobile applications and online platforms, can provide farmers with access to real-time information on weather conditions, market prices, and best practices for agroforestry management. These tools can facilitate knowledge sharing and collaboration among farmers, researchers, and extension services.

(b) ***Remote Sensing:*** Remote sensing technologies, such as satellite imagery and drone-based monitoring, can offer valuable insights into the health and productivity of agroforestry systems. These tools can be used to assess crop and tree growth, monitor soil moisture levels, and detect early signs of pest infestations or nutrient deficiencies. Remote sensing data can also support landscape-level planning and management, helping to identify suitable areas for agroforestry expansion.

(c) ***Precision Agriculture:*** Precision agriculture tools, including GPS-guided equipment and sensor-based technologies, can optimize resource use within agroforestry systems. These tools enable precise application of inputs such as water, fertilizers, and pesticides, reducing waste and minimizing environmental impacts. Precision agriculture can also enhance the efficiency of planting and harvesting operations, improving overall productivity.

(d) ***Decision Support Systems:*** Decision support systems integrating climate data and predictive modeling can facilitate real-time adaptation strategies, empowering farmers with actionable insights to mitigate production risks and optimize resource allocation. These systems can help farmers anticipate and respond to climatic variations, manage water resources more effectively, and plan for contingencies such as pest outbreaks or market fluctuations.

(e) ***Climate Data and Predictive Modeling:*** Integrating climate data and predictive modeling into agroforestry management can help farmers adapt to changing environmental conditions. These tools can provide forecasts of weather patterns, identify potential climate-related risks, and suggest adaptive management practices to enhance system resilience.

By focusing on these priority research areas and leveraging advanced technologies, we can address agroforestry systems' complex challenges and unlock their full potential for sustainable agricultural development.

Future Prospects and Opportunities

Looking ahead, agroforestry's future is imbued with transformative potential, offering a pathway towards sustainable agricultural intense-fication, rural development, and climate resilience.

Potential for Scaling Up Agroforestry Practices

The scalability of agroforestry practices hinges upon fostering enabling environments that promote inclusive participation, knowledge exchange, and adaptive management. By creating supportive frameworks and policies, we can facilitate the widespread adoption of agroforestry systems, ensuring their benefits reach diverse agricultural landscapes.

(a) **Enabling Environments**: Strengthening multi-stakeholder partnerships is crucial for scaling agroforestry practices. Collaborative efforts involving government agencies, research institutions, civil society organizations, and private sector entities can drive technology transfer and scaling initiatives. These partnerships can support infrastructure development, provide financial incentives, and create favorable policy conditions that encourage adopting agroforestry practices.

(b) **Inclusive Participation and Knowledge Exchange**: Encouraging inclusive participation and knowledge exchange is essential for the success of agroforestry initiatives. Farmer-centered approaches, prioritizing local knowledge systems and participatory decision-making, empower communities to co-design and implement resilient agroforestry models tailored to their socio-cultural and environmental contexts. Training programmes, field schools, and demonstration sites can facilitate sharing of best practices and innovations, enhancing farmers' capacity to adopt and adapt agroforestry systems.

(c) **Adaptive Management**: Promoting adaptive management practices ensures that agroforestry systems remain resilient and responsive to changing conditions. This involves continuous monitoring, evaluation, and learning to refine practices and address emerging challenges. Farmers can leverage digital tools and data-driven insights to optimize their management strategies, improving agroforestry systems' overall effectiveness and sustainability.

(d) **Economic Incentives**: Providing economic incentives such as subsidies, grants, and market access can motivate farmers to adopt agroforestry practices. Financial support for initial investments in tree planting and maintenance can lower the barriers to entry, while access to markets for timber, non-timber forest products, and agroforestry crops can enhance farmers' income and economic stability.

(e) **Climate Resilience**: Scaling up agroforestry practices can significantly contribute to climate resilience by improving soil health, enhancing water management, and sequestering carbon. Agroforestry systems can buffer against climate shocks, such as droughts and floods, by creating more diverse and stable agricultural landscapes. By integrating trees into farming systems, agroforestry enhances biodiversity, providing habitat for wildlife and promoting ecological balance.

In short, agroforestry's future prospects and opportunities are vast and promising. We can scale agroforestry systems to achieve sustainable agricultural intensification, rural development, and climate resilience by fostering enabling environments, promoting inclusive participation, and adopting adaptive management practices. These efforts will enhance smallholder farmers' livelihoods and contribute to broader environmental and socio-economic goals.

Integrating Agroforestry with National Development Goals

Integrating agroforestry into national development agendas aligns with the overarching goals of poverty alleviation, food security, and environmental sustainability. Policy coherence across sectors- agriculture, forestry, environment, and finance- is essential to mainstreaming agroforestry within national strategies and frameworks. Incentivizing sustainable land use practices through policy instruments, fiscal incentives, and market-based mechanisms stimulates private sector investments in agroforestry value chains, unlocking new avenues for economic growth and rural transformation.

Furthermore, leveraging international partnerships and South-South cooperation facilitates knowledge sharing, technology transfer, and capacity-building initiatives that transcend geopolitical boundaries. Collaborative research platforms and global networks amplify the collective impact of agroforestry innovations, catalyzing transformative change at local, regional, and global scales.

In Summary, navigating the complexities of integrated farming systems, particularly agroforestry, requires a paradigm shift towards inclusive and sustainable agricultural development. Overcoming technical, socio-economic, and environmental barriers demands synergistic efforts across disciplines, sectors, and scales. By harnessing the transformative potential of agroforestry, we can forge resilient farming landscapes that sustainably meet the evolving needs of present and future generations. Embracing innovation, fostering partnerships, and prioritizing policy coherence are imperative to realizing the full promise of integrated farming systems in fostering a prosperous and resilient agricultural future.

Embracing the Future of Agroforestry

As we reflect on the transformative potential of integrated farming systems, particularly agroforestry, it becomes evident that sustainable agricultural development hinges upon innovative approaches that harmonize ecological integrity, economic viability, and social inclusivity. This concluding section encapsulates the essence of our collective journey towards a resilient and prosperous agricultural future, underscoring key insights, envisioning a sustainable vision for agroforestry in India, and issuing a compelling call to action for stakeholders across sectors.

Summary of Key Points

Throughout this discourse, we have elucidated the multifaceted benefits of integrated farming systems, emphasizing agroforestry as a linchpin in sustainable agricultural intensification. Integrated approaches, encompassing the strategic integration of trees, crops, and livestock, enhance agricultural productivity, bolster ecosystem resilience, mitigate climate risks, and foster socio-economic well-being. By optimizing resource use efficiency and minimizing environmental footprints, agroforestry models exemplify a paradigm shift towards regenerative agriculture that aligns with global sustainability imperatives.

Vision for the Future of Agroforestry in India

Looking forward, the future of agroforestry in India is anchored in a bold vision of sustainable rural development, food security, and environmental stewardship. Embracing agroforestry as a cornerstone of national agricultural policies holds immense potential to rejuvenate degraded lands, enhance water-use efficiency, and mitigate greenhouse gas emissions. By mainstreaming agroforestry within national development agendas, India can harness its diverse agroecological zones to foster climate-resilient farming systems that sustainably meet the nutritional needs of a burgeoning population.

The vision entails fostering inclusive growth, empowering smallholder farmers, and revitalizing rural economies through diversified income streams derived from tree products, non-timber forest produce, and eco-tourism initiatives. Furthermore, integrating traditional knowledge systems with modern agronomic practices enhances adaptive capacity and fosters community resilience amidst changing climatic conditions. By nurturing vibrant agroforestry value chains, India can be a global leader in sustainable agriculture, balancing agricultural productivity with environmental conservation and socio-economic equity.

Call to Action for Stakeholders

Realizing this transformative vision necessitates collective action and unwavering commitment from stakeholders across government, academia, civil society, and the private sector. It calls for a paradigmatic shift in policy frameworks, institutional arrangements, and investment priorities to prioritize sustainable land use practices and agroecological resilience.

Policy Imperatives

First and foremost, policymakers must prioritize agroforestry within national agricultural policies, ensuring coherence across the forestry, environment, and rural development sectors. Incentivizing agroforestry investments through fiscal incentives, subsidy schemes, and credit facilities encourages farmers to adopt sustainable land management practices that enhance productivity and resilience.

Research and Innovation

Investing in research and innovation is paramount to advancing agroforestry science, technology, and extension services. Collaborative research endeavors, supported by robust data analytics and digital technologies, generate evidence-based solutions to optimize tree-crop-livestock interactions, improve soil fertility, and mitigate pest and disease pressures. By harnessing indigenous knowledge systems and participatory research methodologies, stakeholders co-create contextually relevant agroforestry models that empower farming communities and enhance adaptive capacity.

Capacity Building and Knowledge Exchange

Empowering stakeholders with technical skills, knowledge, and extension services is essential to catalyze the adoption of agroforestry practices at scale. Strengthening farmer cooperatives, producer organizations, and women's self-help groups fosters inclusive participation and collective action towards sustainable agricultural intensification. Furthermore, promoting farmer-to-farmer knowledge exchange networks and establishing demonstration sites showcase best practices, instilling confidence and fostering peer learning among farmers.

Market Integration and Value Chain Development

Facilitating market access for agroforestry products through robust market linkages, certification schemes, and fair trade practices enhances economic opportunities for smallholder farmers and forest dwellers. Developing sustainable agroforestry value chains for timber, fruits, nuts, medicinal plants, and bio-products diversifies income streams, reduces dependency on monoculture crops, and enhances rural livelihoods. By fostering market-based incentives and public-private partnerships, stakeholders unlock the economic potential of agroforestry while safeguarding environmental integrity.

Community Engagement and Empowerment

Central to the success of agroforestry initiatives is inclusive governance and participatory decision-making that empowers local communities as stewards of natural resources. Strengthening community forest rights, promoting gender equity, and fostering social cohesion cultivate a sense of ownership and responsibility towards sustainable land management practices. Empowering youth through vocational training and entrepreneurial opportunities cultivates a new generation of agroforestry champions who drive innovation and advocate for sustainable development.

Global Collaboration and Knowledge Sharing

Lastly, fostering international partnerships and South-South cooperation enhances knowledge sharing, technology transfer, and capacity-building initiatives that transcend geographical boundaries. Collaborative platforms, such as research consortia, global networks, and multi-stakeholder dialogues, amplify the impact of agroforestry innovations, catalyzing transformative change in agricultural landscapes worldwide.

Conclusion

The journey towards sustainable agricultural futures through agroforestry is both a moral imperative and an economic opportunity. As global challenges like climate change, biodiversity loss, and food insecurity intensify, agroforestry offers a viable solution that integrates ecological resilience, economic prosperity, and social equity. For India, a nation with a rich heritage of traditional agricultural practices and a burgeoning need for sustainable development, agroforestry represents a pathway to a sustainable agricultural renaissance that nourishes both people and the planet.

Agroforestry, by design, harnesses the synergy between trees, crops, and livestock, creating more productive, sustainable, and resilient systems to climatic shocks. This holistic approach enhances biodiversity and soil health and provides multiple income streams for farmers, thereby reducing economic vulnerability. The economic benefits of agroforestry are significant, including increased crop yields, timber production, and non-timber forest products, all of which contribute to improved livelihoods and poverty alleviation. Moreover, agroforestry systems can be crucial in carbon sequestration, mitigating climate change and contributing to global efforts to limit temperature rise.

India's commitment to agroforestry can be seen in its National Agroforestry Policy, which aims to integrate trees into farming systems on a large scale. This policy underscores the importance of research and development, capacity building, and financial support to farmers adopting agroforestry practices. By fostering an enabling environment, the policy aims to enhance the economic viability of agroforestry, making it a lucrative option for farmers across the country.

However, realizing the full potential of agroforestry requires more than just policy support. It necessitates a paradigm shift in how we view and manage agricultural landscapes. Embracing innovation is pivotal. Digital technologies, such as Geographic Information Systems (GIS) and remote sensing, can revolutionize the planning and monitoring of agroforestry systems. Precision agriculture tools can optimize resource use, improve crop and tree health, and enhance productivity. Integrating these technologies with traditional knowledge systems can create robust, adaptive, and sustainable agroforestry models.

Fostering partnerships is equally important. Multi-stakeholder collaboration, including government agencies, research institutions, civil society organizations, and the private sector, can accelerate technology transfer, enhance knowledge exchange, and scale successful agroforestry practices. Public-private partnerships can drive agroforestry investment, ensuring farmers have access to quality planting material, technical advice, and markets for their produce. Moreover, involving local communities in decision-making ensures that agroforestry initiatives are tailored to different regions' socio-cultural and environmental contexts, enhancing their acceptance and success.

Prioritizing inclusive development is crucial to ensuring that the benefits of agroforestry are equitably distributed. Special attention must be given to marginal and smallholder farmers, who are often the most vulnerable to climate change and economic shocks. Providing them with access to resources, training, and markets can empower them to adopt and benefit from agroforestry practices. Gender-inclusive approaches are also essential, as women play a vital role in agriculture and natural resource management. Ensuring their participation and leadership in agroforestry initiatives can enhance the effectiveness and sustainability of these efforts.

Agroforestry also has the potential to safeguard biodiversity. By creating diverse and multifunctional landscapes, agroforestry systems can provide habitats for various species, thereby enhancing biodiversity conservation. This, in turn, supports ecosystem services such as pollination, pest control, and water regulation, which are critical for agricultural productivity and resilience. By integrating biodiversity conservation into agricultural practices, agroforestry contributes to the broader goal of sustainable land management.

As stewards of our shared future, we must heed the call to action and forge a path towards a sustainable and resilient agricultural landscape. Agroforestry thrives as a beacon of hope and prosperity, rooted in the wisdom of nature and the aspirations of humanity. By embracing innovation, fostering partnerships, and prioritizing inclusive development, we can cultivate a legacy of sustainability. This legacy will leave a flourishing planet for posterity and ensure food security for all.

In conclusion, the transformative potential of agroforestry lies in its ability to create win-win scenarios for people and the planet. It offers a sustainable way to meet a growing population's food, fiber, and fuel needs while conserving natural resources and mitigating climate change. Integrating trees into agricultural systems can enhance productivity, build resilience, and create sustainable livelihoods. The journey towards sustainable agricultural futures through agroforestry is not just a possibility but a necessity. Together, we can cultivate a legacy of sustainability, leaving a flourishing planet for future generations.

References

[1] Altieri, M.A. and Nicholls, C.I. (2004). Biodiversity and pest management in agroecosystems (2nd ed.). CRC Press.

[2] FAO (2019). State of the World's Forests 2018: Forest Pathways to Sustainable Development. Food and Agriculture Organization of the United Nations.

[3] Franzel, S., Coe, R. and Cooper, P. (2001). Assessing the adoption potential of agroforestry practices in sub-Saharan Africa. *Agricultural Systems,* 69(1–2), 37–62.

[4] Garrity, D.P., Akinnifesi, F.K., Ajayi, O.C., Weldesemayat, S.G., Mowo, J.G. and Kalinganire, A. (2010). Evergreen agriculture: A robust approach to sustainable food security in Africa. Food Security, 2(3), 197–214.

[5] Harvey, C.A., Chacón, M., Donatti, C.I., Garen, E., Hannah, L., Andrade, A. and Bede, L. (2014). Climate-smart landscapes: Opportunities and challenges for integrating adaptation and mitigation in tropical agriculture. Conservation Letters, 7(2), 77–90.

[6] Hosmani, M.P. (2014). Agroforestry Systems in India: Livelihood Security & Ecosystem Services. Narosa Publishing House.

[7] Huxley, P. (1999). Tropical Agroforestry. Blackwell Science.

[8] Kang, B.T., Reynolds, L. and Atta-Krah, A.N. (1990). Alley Farming. International Institute of Tropical Agriculture.

[9] Kumar, B. M. and Nair, P. K. R. (2011). The enigma of tropical home gardens. Agroforestry Systems, 82(3), 173–184.

[10] Leakey, R.R.B. (2012). Living with the trees of life—Towards the transformation of tropical agriculture. Agronomy for Sustainable Development, 32(2), 491–507.

[11] Leakey, R.R.B. (2014). Agroforestry for Biodiversity and Ecosystem Services. Springer.

[12] Mbow, C., Smith, P., Skole, D., Duguma, L. and Bustamante, M. (2014). Achieving mitigation and adaptation to climate change through sustainable agroforestry practices in Africa. Current Opinion in Environmental Sustainability, 6, 8–14.

[13] Ministry of Agriculture, Govt of India. (2016). National Agroforestry Policy 2014. Department of Agriculture and Cooperation.

[14] Ministry of Agriculture, Govt of India. (2017). Pradhan Mantri Krishi Sinchayee Yojana (PMKSY)—Guidelines. Department of Agriculture, Cooperation & Farmers Welfare.

[15] Ministry of Agriculture, Govt of India. (2018). Mission for Integrated Development of Horticulture (MIDH)—Operational Guidelines. Department of Agriculture, Cooperation & Farmers Welfare.

[16] Ministry of Agriculture, Govt of India. (2019). National Policy for Farmers 2007. Department of Agriculture and Cooperation.

[17] Ministry of Agriculture, Govt of India. (2020). National Mission for Sustainable Agriculture (NMSA)—Guidelines. Department of Agriculture, Cooperation & Farmers Welfare.

[18] Ministry of Agriculture, Govt of India. (2021). Paramparagat Krishi Vikas Yojana (PKVY)—Operational Guidelines. Department of Agriculture, Cooperation & Farmers Welfare.

[19] Ministry of Agriculture, Govt of India. (2022). National Livestock Mission (NLM)—Revised Guidelines. Department of Animal Husbandry, Dairying & Fisheries.

[20] Ministry of Agriculture, Govt of India. (2023). Agroforestry Mission—Implementation Framework. Department of Agriculture, Cooperation & Farmers Welfare.

[21] Ministry of Agriculture, Govt. of India. (2023). Soil Health Management—Policy and Guidelines. Department of Agriculture, Cooperation & Farmers Welfare.

[22] Ministry of Agriculture, Govt. of India. (2023). Sustainable Agriculture Practices—Policy Overview. Department of Agriculture, Cooperation & Farmers Welfare.

[23] Montagnini, F. and Nair, P.K.R. (2004). Carbon sequestration: An underexploited environmental benefit of agroforestry systems. Agroforestry Systems, 61–62(1–3), 281–295.

[24] Nair, P.K.R. (1993). An introduction to agroforestry. Kluwer Academic Publishers.

[25] Nair, P.K.R. (2012). Climate change mitigation: A low-hanging fruit of agroforestry. Environmental Science & Policy, 16, 88–95.

[26] Place, F., Barrett, C.B., Freeman, H.A., Ramisch, J.J. and Vanlauwe, B. (2017). Prospects for integrated soil fertility management using organic and inorganic inputs: Evidence from smallholder African agricultural systems. Food Security, 9(1), 115–127.

[27] Pretty, J. (2008). Agricultural sustainability: Concepts, principles and evidence. Philosophical Transactions of the Royal Society B: Biological Sciences, 363(1491), 447–465.

[28] Rockström, J., Williams, J., Daily, G., Noble, A., Matthews, N., Gordon, L., ... and Smith, J. (2017). Sustainable intensification of agriculture for human prosperity and global sustainability. Ambio, 46(1), 4–17.

[29] Sanchez, P.A., Palm, C.A. and Buol, S.W. (2003). Fertility capability soil classification: A tool to help assess soil quality in the tropics. Geoderma, 114(3–4), 157–185.

[30] Shiferaw, B., Prasanna, B. M., Hellin, J. and Bänziger, M. (2011). Crops that feed the world 6. Past successes and future challenges to the role played by maize in global food security. Food Security, 3(3), 307–327.

[31] Sinclair, F.L. (2004). Agroforestry Science, Policy and Practice. Springer.

[32] Smith, P., Martino, D., Cai, Z., Gwary, D., Janzen, H., Kumar, P., ... & Smith, J. (2007). Agriculture. In Climate Change 2007: Mitigation. Contribution of Working Group III to the Fourth Assessment Report of the Intergovernmental Panel on Climate Change. Cambridge University Press.

[33] Smith, S.E. and Read, D.J. (2008). Mycorrhizal Symbiosis (3rd ed.). Academic Press.

[34] Swift, M.J., *et al.* (2004). Biodiversity and Ecosystem Services in Agricultural Landscapes. Island Press.

[35] Thomas, R.J., Ayarza, M.A. and Amézketa, E. (2014). Soil fertility management in agroecosystems: Building new paradigms from traditional knowledge. Sustainability, 6(1), 370–398.

[36] Tsonkova, P., Böhm, C., Quinkenstein, A., Freese, D. and Schaller, E. (2014). Energetic evaluation of short rotation coppice systems for the production of woody biomass for energy in the European Union. Renewable Energy, 62, 97–105.

[37] van der Meer, H.G., *et al.* (2002). Soil Fertility in Sustainable Low-Input Farming Systems. Wageningen Academic Publishers.

[38] Verchot, L.V., Van Noordwijk, M., Kandji, S., Tomich, T., Ong, C.K., Albrecht, A. and Mackensen, J. (2007). Climate change: linking adaptation and mitigation through agroforestry. Mitigation and Adaptation Strategies for Global Change, 12(5), 901–918.

[39] Wani, S.P., Chander, G., Sahrawat, K.L., Raju, K.V., Palanisami, K., Marimuthu, S. and Pavani, M. (2008). Sustaining crop-livestock production systems through integrated nutrient management. SAT eJournal, 5(1), 1–4.

[40] Wollenberg, E., Richards, M., Smith, P., Havlík, P., Obersteiner, M., Tubiello, F.N., ... and Obersteiner, M. (2016). Reducing emissions from agriculture to meet the 2°C target. Global Change Biology, 22(12), 3859–3864.

[41] World Agroforestry Centre (ICRAF). (2009). Trees for Food Security. World Agroforestry Centre.

[42] Zomer, R.J., Trabucco, A., Coe, R. and Place, F. (2009). Trees on farm: Analysis of global extent and geographical patterns of agroforestry. *ICRAF Working* Paper No. 89.

Soil Health

Odemari Mbuya[1]

Introduction

Soil is the unconsolidated uppermost layer of the Earth's crust. It is a complex mixture of minerals, organic matter, air, water, and organisms that support life. It is an intersection of the lithosphere (minerals), atmosphere (air), hydroshere (water) and biosphere (organisms) called pedosphere (soil). When living organisms die (plus animal excrete/manure) constitute soil organic matter. Soils perform important functions to sustain plant and animal life, regulate water flow, filter and buffer pollutants, cycle nutrients, provide physical stability for engineering, and regulate atmospheric temperature. A typical soil consists of approximately 45% mineral, 5% organic matter, 2% water, and 25% air. These percentages may vary depending on the type and condition of soil. For example, sandy soils will have less organic matter, whereas wet soils will have more water and less air. A well weathered soil has a profile (vertical cross section) that is characterized by horizons (layers parallel to the soil surface whose physical, chemical, and biological characteristics differ from the layers above and beneath) that are distinguishable from one another.

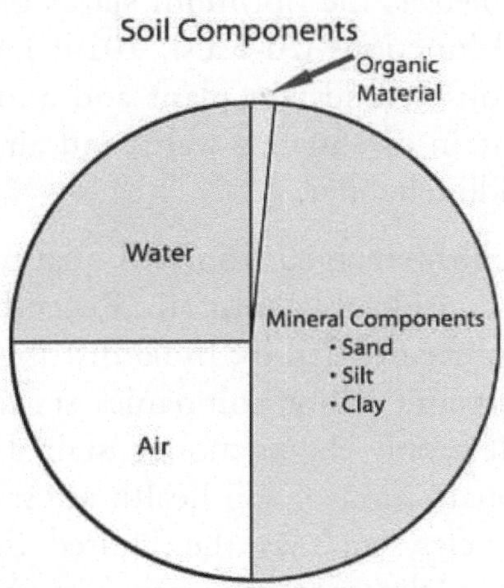

Figure 1: Components of Typical Soil

[1] Center for Water Resources, Florida Agricultural and Mechanical University (USA).

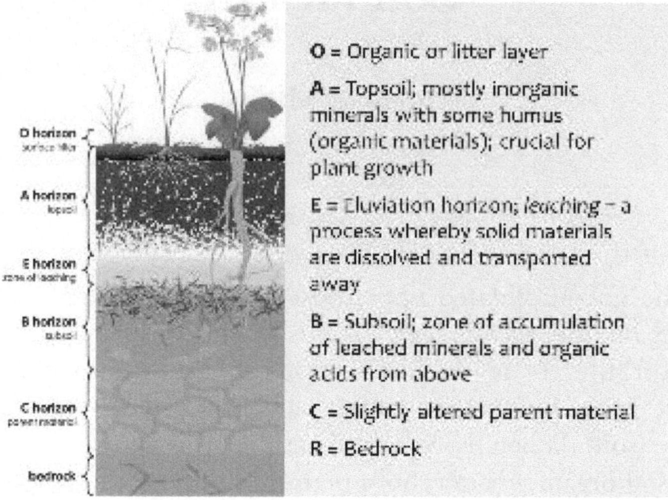

Figure 2: Soil and Soil Profile

Soil health is the continued capacity of soil to function as a vital living ecosystem that sustains plants, animals, and humans, and connects agricultural and soil science to policy, stakeholder needs and sustainable supply chain management. Soil health is essential for agriculture, and crucial to many other ecosystem services. Quantifying soil health is still dominated by chemical indicators, despite growing appreciation of the importance of soil biodiversity, due to limited functional knowledge and lack of effective methods. In general, soil health, as a measure of soil functions, can be defined as the optimum status of the soil's biological, physical and chemical functions (Al-Kaisi, 2014; Lehmann *et al.*, 2020). This means healthy soils can sustain plant and animal productivity and soil biodiversity, maintain or enhance water and air quality, and support human health and wildlife habitat.

The concept of soil health emerged from soil quality in the 1990s, and it was initially met with considerable criticism. Recently, policy makers have embraced the concept, exemplified by India distributing soil health cards to 100 million farmers and major companies starting programs on soil health to manage their supply chains more sustainably. The terminology, concept, and operationalization of soil health are still evolving. It is now defined by most agencies, such as the United States Department of Agriculture (USDA), as "the continued capacity of soil to function as a vital living ecosystem that sustains plants, animals, and humans" (https://www.nrcs.usda.gov/wps/portal/nrcs/main/soils/health/). Several other related concepts exist, including soil fertility, soil quality, and soil security, which also emphasize the role or functioning of soil in society, ecosystems and/or agriculture. On 26 February 1937, the 32nd President of the USA, Franklin D. Roosevelt, wrote to all State Governors that, *"The nation that destroys its soil destroys itself"*. This emphasizes the importance of soils to society and civilization as his plea related directly to the extreme climate events occurring in the USA at that period; it also demonstrated political leadership in soil management and conservation to address the risk of its wholesale decline and degradation.

Achieving healthy soil with the right balance of minerals, organic matter, water, air, and microorganisms isn't easy, that's why you need help from professionals. In intensive modern agriculture, the role of science, engineering and technology in maintaining a health soil can neither be overlooked nor ignored.

An Ecosystem

An ecosystem is a geographic area where living organisms (biotic) and non-living (abiotic) factors interact with each other. An ecosystem can be of different sizes and types, such as marine, aquatic, or terrestrial. An ecosystem includes all the organisms and the physical environment in a specific area. In other words, an ecosystem is a unit of the biosphere where matter and energy are conserved.

Ecosystem Functions and Services

"The most unique feature of Earth is the existence of life, and the most extraordinary feature of life is its diversity. Approximately 9 million types of plants, animals, protists and fungi inhabit the Earth. So, too, do 7 billion people. Two decades ago, at the first Earth Summit, the vast majority of the world's nations declared that human actions were dismantling the Earth's ecosystems, eliminating genes, species and biological traits at an alarming rate. This observation led to the question of how such loss of biological diversity will alter the functioning of ecosystems and their ability to provide society with the goods and services needed to prosper." – Bradley J. Cardinale

Processes carried out by ecosystems provide a variety of products and services for humanity and can be divided into two primary categories; i) ecosystem functions and ii) ecosystem services. Ecosystem functions can be defined by "the ecological processes that control the fluxes of energy, nutrients and organic matter through an environment". Functions within the ecosystem that help maintain the Earth's natural balance such as primary production, decomposition of dead matter and nutrient recycling. Ecosystem services are processes that occur within an ecosystem provide benefits to humanity. Services such as food, water and oxygen. These services can be splits into provisioning services which provide humans and other species with tangible goods such as food and water; and regulating services that help regulate the Earth's environment and atmosphere. Regulating services include water purification, climate regulation and oxygen production.

Soil as an Ecosystem

Soils are diverse ecosystems comprising of living and nonliving elements that interact in various ways (Sokol *et al.*, 2020; Anikwe and Ife, 2023). Soil is considered an ecosystem because it is a dynamic and diverse natural system that lies at the interface between earth, air, water, and life. Soil contains a multitude of different organisms, which interact and contribute to the global cycles that make all life possible. Soil is a major reservoir of biodiversity and a critical provider of ecosystem services for the sustenance of humanity.

Soil Health and Ecosystem Functions

An ecosystem is a community of living organisms (plants, animals and microbes) in a particular area. Ecosystem functions can be defined by "the ecological processes that control the fluxes of energy, nutrients and organic matter through an environment". Ecosystem processes and functions describe biophysical relationships and exist regardless of whether or not humans benefit. Ecosystem services, on the other hand, only exist if they contribute to human well-being and cannot be defined independently (Sekercioglu, 2010).

Soil Health and Ecosystem Services

Ecosystem services are the ecosystem functions that have direct and indirect contributions and benefits to humans, their wellbeing and life quality. Such benefits include the purification of water and air, detoxification and decomposition of wastes, regulation of climate, regeneration of soil fertility, providing safe food, cultural aspects (meditation, reducing stress and anxiety), aesthetics (leisure, scenery beauty), and maintenance of biodiversity. In a nutshell, ecosystem services are benefits provided to human societies by natural ecosystems. Maintaining natural ecosystems (e.g. health soils) enhances ecosystem services.

Soils provide critical ecosystem services, especially for sustaining ecosystems and growing food crops, but soil erosion and degradation are serious problems worldwide (Sekercioglu, 2010). Higher biodiversity usually increases ecosystem efficiency and productivity, stabilizes overall ecosystem functioning, and makes ecosystems more resistant to perturbations. Mobile linked animal species provide critical ecosystem functions and increase ecosystem resilience by connecting habitats and ecosystems through their movements. Their services include pollination, seed dispersal, nutrient deposition, pest control, and scavenging. Thousands of species that are the components of ecosystems harbor unique chemicals and pharmaceuticals that can save people's lives, but traditional knowledge of medicinal plants is disappearing and many potentially valuable species are threatened with extinction. Increasing habitat loss, climate change, settlement of wild areas, and wildlife consumption facilitate the transition of diseases of animals to humans, and other ecosystem alterations are increasing the prevalence of other diseases. Valuation of ecosystem services and tradeoffs helps integrate these services into public decision-making and can ensure the continuity of ecosystems that provide the services.

Soils provide multiple ecosystem services, and as such, soil health management in support of sustainability must consider three points; i) that enhancing many soil ecosystem services requires multi-functional management, ii) that managing soil to improve one service can have positive (synergistic) or negative effects (tradeoffs) on another service, and iii) that soil health management should sustain soil services over the long term. As an example, four main soil ecosystem services are highlighted during soil health management, i) sustainable plant production, ii) water quality control, iii) human health advancement, and iv) climate change mitigation.

Based on available scientific evidence, we are certain that ecosystem services are essential to civilization (Gretchen *et al.*, 2018). Ecosystem services operate on such a grand scale and in such intricate and little-explored wats that most could not be replaced by technology. Human activities are already impairing the flow of ecosystem services on a large scale. If current trends continue, humanity will dramatically alter virtually all of Earth's remaining natural ecosystems within a few decades. Many of the human activities that modify or destroy natural ecosystems may cause deterioration of ecological services whose value, in the long term, dwarfs the short-term economical benefits society gains from those activities. Considered globally, very large number of species and populations are required to sustain ecosystem services. The functioning of many ecosystems could be restored if appropriate actions were taken in time. Land use and development policies should strive to achieve a balance between sustaining vital ecosystem and pursuing the worthy short-term goals of economic development (Gretchen *et al.*, 1997).

Soil Health and Human Health

Human health depends to a great extent on soil health, including and going beyond the obvious connection between soil and human health through crop production. For example, soils with greater micronutrient availability are related to lower malnutrition (Barrett and Bevis, 2015), while soils with high organic matter improves the nutritional value of crops (Wood *et al.*, 2018). Nutritional value of crops can also depend on robust soil biodiversity (Wall *et al.*, 2015), which can enhance micronutrient bioavailability to crops (Jacoby *et al.*, 2017) and suppress soil-borne plant disease (Schlatter *et al.*, 2017), as well as taste, food storage and preparation (Rillig *et al.*, 2018).

Soils can negatively impact human health. For examples, soil pollutants can contaminate produce through direct contact or dust, suspension, or rain splash. Some compounds, such as arsenic (Oliver and Gregory, 2015) as with most inorganic pollutants, can also be taken up through the root system and accumulate in grain or fruit. In addition to abiotic contaminants, soils can contain pathogenic fungi that produce mycotoxins, contaminating plant products and causing acute and chronic diseases (Hussein and Brasel, 2001) in animals and humans. Furthermore, soils are also the source of parasitic worms (helminthiasis) that can live for years in the human gastrointestinal tract, cause malnutrition, and result in stunted development (Bethony *et al.*, 2006). In India, a direct linkage between the health and growth of children, and soil micronutrients has been explored. Specifically, regions in India with low levels on zinc and iron in the soil have children with growth deficiencies, diarrheal issues, lower IQs, and immune compromises.

Although soil hosts pathogens, it has also historically been the source of organisms that produce antibiotics used in the medical industry, such as streptomycin (Schatz *et al.*, 1944). Most of the soil microbiome remains to be identified, and important discoveries for human medical applications may still be made (Ling *et al.*, 2015). Quantifying and managing soil biodiversity, part of the goals of soil health management, is needed to arrest extinction of microbial species (Veresoglou *et al.*, 2015) and preserving opportunities for future bioprospecting.

Soil Heath and Sustainable Crop Production

Crop production, the main goal of intensive agriculture, is an important focus of soil health management (Jian *et al.*, 2020; Karlen *et al.*, 2019), as it affects water use and quality, human health, animal health, climate and biodiversity. A foundation of soil health, though, is the recognition that managing nutrient availability alone, such as through the use of agrochemicals (mainly fertilizers), is not sufficient for optimizing plant growth (Bünemann, 2018). Furthermore, there is increased recognition that some management practices used in intensive agriculture to increase total plant production are detrimental to soil health (Norris and Congreves, 2018). For example, rooting depth, critical in plant production, depends to a large extent on soil structure, which is determined in part by organic matter content (O' Dell and Claassen, 2006) and soil preparation (Congreves *et al.*, 2015). Tillage can negatively impact soil structure through soil compaction (Hamza and Anderson, 2005), and the use solely of inorganic fertilizers (as opposed to organic rich fertilizers such as compost and manure, or the use of cover crops) is often not sufficient to restore or retain adequate levels of soil organic matter (Jenkinson, 1991). Focusing on soil health will therefore expand soil management from a reliance on inorganic fertilizers to employing organic amendments and crop residue return, reducing mechanical impact by tillage, increasing plant diversity in both time and space, or reducing erosion with contour ploughing (ploughing along elevation contours) or grass strips (Karlen *et al.*, 2019; O' Dell and Claassen, 2006; Congreves *et al.*, 2015).

In addition to managing physicochemical soil properties for plant production, soil health considers the interactions between plants and soil microbial communities around roots, which can promote or reduce plant growth (Berendsen *et al.,* 2012). Promoting a soil microbiome for high plant production requires management of microbial abundance and activity, community composition and specific functions (Chaparro *et al.*, 2012; Bonanomi *et al.*, 2018). For example, organic amendments (such as compost) can foster increased resilience to plant pathogens through promotion of beneficial microorganisms (Bonanomi *et al.*, 2018). In many cases, higher organic matter contents through higher amendments or reduced tillage increase biodiversity that is expected to improve crop resilience (Chen *et al.*, 2020). However, there are exceptions to these trends, as for example reducing tillage may reduce crop yields in some instances (Pittelkow *et al.*, 2015) with follow-on reductions of soil organic carbon (Ogle *et al.*, 2012).

Soil Health and Water Quality

Soils can be a source and/or sink of pollutants (Zimnicki *et al.*, 2020) as rainwater and snowmelt moves through it. These pollutants include herbicides, pesticides, heavy metals, antibiotics, hormones, microplastics, pathogens, polycyclic aromatic hydrocarbons (PAH), per- and poly-fluoroalkyl substances (PFAS) (Evans *et al.*, 2019). Moreover, nutrient pollution from agricultural fertilizer use is a global problem, leading to eutrophication and/or anoxia of waterways, promoting harmful algal blooms, and negatively impacting drinking water quality Carpenter *et al.*, 1998). Thus, there is a trade-off between soil management to support crop growth and water quality, which requires careful consideration and multiple management strategies.

Managing soil health to promote good water quality includes retaining pollutants and others in the soil, buffering against them, and biotically transforming them. Increasing soil organic matter will retain heavy metals and organic toxins, some of which show nearly irreversible adsorption to organic matter (Lamichhane *et al.*, 2016). Using buffer zones, such as vegetative filter strips near agricultural areas or constructed wetlands, can slow the migration of nitrate, phosphate or pesticide contamination to water (Tournebize *et al.*, 2017). Soil biota can transform organic pollutants, such as the common hydrocarbon toluene, to harmless compounds (Hanson *et al.*, 1999). Therefore, both organic matter content and microbial activity, key properties of soil health, improve the quality of the water that is draining soil.

Soil health of urban soils have not yet received sufficient recognition (Li *et al.*, 2018), but can contain an even wider range of contaminants than agricultural soils, and many urban soils have also been modified to an extent that water can drain either very quickly or not at all (Laurenson *et al.*, 2013). Soil health management in urban soils must therefore balance eliminating surface runoff against retaining water and pollutants by reduced drainage. A combination of managing physical retention with biological transformation of pollutants through high soil biodiversity (Wagg *et al.*, 2014) is the goal of bioretention (Laurenson *et al.*, 2013) and constructed soils (Kadam *et al.*, 2008) to provide clean drinking water.

Soil Health and Climate Change

Soil management can mitigate or exacerbate climate change and its effects on other soil ecosystem services such as water quality or plant production (Lal, 2004; Paustian *et al.*, 2016). For example, climate change mitigation strategies, such as sequestering carbon in soil as organic matter, can benefit agriculture by improving crop productivity and resilience to drought and flooding (Lal, 2004). Increased soil organic matter contents can be achieved by increasing the use of organic fertilizers or soil amendments, as well as by reducing tillage (Karlen *et al.*, 2019) to increase aggregation and control microbial mineralization to carbon dioxide, which can also promote plant growth. However, there are trade-offs between managing soil health for climate change versus for food production. For instance, use of nitrogen fertilizers, which are commonly used to increase crop production, can lead to increased emissions of nitrous oxide, which is a powerful greenhouse gas (Paustian *et al.*, 2016). These examples highlight the difficulty in balancing the various uses of soils, and why it is important to provide context and goals for soil health management.

Soil Health Assessment for Crop Production, Climate Change and Water Quality

Soil health assessments for plant production often include total organic carbon, plant-available nutrients (fertility), pH, cation exchange capacity (CEC), electrical conductivity (EC), ground penetration resistance, N mineralization, and microbial biomass. A smaller number of these assessment (tests) include soil aggregation (soil structure), water storage, and organic carbon (OC) fractions.

Managing soil health for climate change mitigation should include testing similar parameters, with a small portion of tests already examining soil nitrogen forms that should be adapted to provide information about potential greenhouse gas emissions including nitrous oxide. Soil health assessments relevant for water quality should include microbial biomass and activity, mobile nutrients, heavy metal toxins, and total organic carbon already part of many soil health testing schemes, yet should also encompass aggregation and infiltration that are only occasionally included. Many of these indicators should also be used in soil health assessment for human health.

In total, more than two thirds of soil health test frameworks currently include the traditional quantification of soil organic matter, pH, and plant-available phosphorus and potassium, and more than half include water storage and bulk density (Bünemann, 2018). A third of tests also recommend measurements of soil respiration, microbial biomass or nitrogen mineralization to characterize biological properties, as well as structural stability (Bünemann, 2018). Chemical indicators make up at least 40% of the indicators in 90% of the soil health assessment schemes, underscoring the continued importance of chemical properties in soil health quantification and the long-standing emphasis on plant production. Indeed, the most advanced analytical schemes currently, such as the Soil Management Assessment Framework, focus on indicators for sustainable crop production (Andrews and Carroll, 2001; Lilburne *et al.*, 2002; Idowu *et al.*, 2009; Cherubin *et al.*, 2016). However, the European Union (EU) Commission recently recommended inclusion of soil biodiversity as one of six indicators of soil health (E.U. Mission Board Soil Health and Food, 2020).

Soil Health Card

The State of Gujarat introduced providing farmers with soil health analysis cards, and farmers were guided to use fertilizer based on deficiency in the soil and crop needs. More importantly, the guidance provided information of crops which can be sustained by the soil. The guidance also provided details of market price of the last three years of suggested crops. This program helped farmers to make informed decisions and developed sustainable agriculture. The program of Soil Health Card was introduced in 2004 when Hon'ble Prime Minister Narendra Modi was Chief Minister of Gujarat, and Dr. Kirit Shelat was Minister of Agriculture of Gujarat.

Climate change has increased soil salinity in coastal regions due to ingress of sea water inland, making soil and water testing key to sustainable agriculture.

Future Perspective of Soil Health

The soil health concept fills an important stakeholder need in sustainable development (E.U. Mission Board Soil Health and Food, 2020) by elevating the recognition for soil's role in modern society and is developing into an attractive and actionable platform for farmers, land managers, municipalities and policy makers. The versatility of the concept allows many stakeholders to adopt soil health and to make it work for their context. By providing an illustrative link to broader sustainability goals that can motivate innovative soil management, soil health meets universal agreement in the eye of the public as a goal to work towards.

Scientists are converging on a definition of soil health, and are developing or refining methods to quantify its various facets, albeit mainly with respect to its crop productivity function and with inadequate consideration of biotic and abiotic diversity. Researchers should embrace soil health as an overarching principle to which to contribute knowledge, rather than as only a property to measure. In this way, soil health could become better established as a scientific field to which many disciplines can contribute, for example by listing their specific discipline's research also under the keyword 'soil health'. Making the soil health concept live up to its potential as a unifying concept that integrates soil functions requires engagement by all involved parties, and particularly a common understanding between stakeholders and scientists.

Because of soils' broad environmental and societal functions, soil health should be legally recognized as a common good. The development of soil health quantification standards should be spearheaded by governmental or intergovernmental organizations such as the Global Soil Partnership. International standards have to be developed for suitable type of indicators, their methodological details (Stott, 2019) and their integration into indices. Such a comprehensive soil health index should then be referenced by local, regional or national jurisdictions and organizations to guide decisions that impact soil and its functions to benefit sustainability goals.

References

[1] Al-Kaisi Mahdi (2014). What is Soil Health and How Can We Improve It? Department of Agronomy, Iowa State University. February 19, 2014. ICM News.

[2] Andrews, S.S. and Carroll, C.R. (2001). Designing a soil quality assessment tool for sustainable agroecosystem management. *Ecol Applic.,* 2001; 11: 1573–1585.

[3] Anikwe, M.A.N. and Ife, K. (2023). The role of soil ecosystem services in the circular bioeconomy. Front. Soil Sci., 02 November 2023 Sec. *Soil Management,* Volume 3, 2023. https://doi.org/10.3389/fsoil.2023.1209100

[4] Barrett, C.B. and Bevis, L.E. (2015). The self-reinforcing feedback between low soil fertility and chronic poverty. *Nature Geosci.* 2015; 8:907–912.

[5] Berendsen, R.L., Pieterse, C.M. and Bakker, P.A. (2012). The rhizosphere microbiome and plant health. *Trends Plant Sci.,* 2012; 17:478–486.

[6] Bethony, J.; Brooker, Simon; Albonico, Marco; Geiger, Stefan M.; Loukas, Alex; Diemert, David and Hotez, Peter J. (2006). Soil-transmitted helminth

infections: ascariasis, trichuriasis, and hookworm. *Lancet.* 2006; 367: 1521–1532.

[7] Bonanomi, G.; Lorito, M,; Vinale, F. and Woo, S.L. (2018). Organic amendments, beneficial microbes, and soil microbiota: Toward a unified framework for disease suppression. *Ann Rev Phytopath.* 2018; 56:1–20.

[8] Bünemann, E.K.; Giulia, Bongiorno; Zhanguo, B.; Rachel, E. Creamer; Gerlinde, De Deyn; Ron de, Goede; Luuk, Fleskens; Violette, Geissen; Thom, W. Kuyper; Paul, Mäder; Mirjam, Pulleman; Wijnand, Sukkel; Jan, Willem van Groenigen and Lijbert, Brussaard (2018). Soil quality-A critical review. *Soil Biol Biochem,* 2018;120: 105–125.

[9] Carpenter, S.R.; Caraco, N.F.; Correll, D.L.; Howarth, R.W.; Sharpley, A.N.; Smith, V.H. (1998). Nonpoint pollution of surface waters with phosphorus and nitrogen. *Ecol Appl.* 1998; 8: 559–568.

[10] Chaparro, J.M.; Sheflin, A.M.; Manter, D.K.; Vivanco, J.M. (2012). Manipulating the soil microbiome to increase soil health and plant fertility. *Biol Fert Soils.* 2012; 48: 489–499.

[11] Chen, X.D., Dunfield, K.E., Fraser, T.D., Wakelin, S.A., Richardson, A.E. and Condron, L.M. (2020). Soil biodiversity and biogeochemical function in managed ecosystems. *Soil Res.* 2020; 58: 1–20.

[12] Cherubin, M.R.; Douglas, L., Karlen, André, L.C. Franco; Carlos, E.P. Cerri; Cássio, A. Tormena and Carlos, C. Cerri (2016). A Soil Management Assessment Framework (SMAF) evaluation of Brazilian sugarcane expansion on soil quality. *Soil Sci Soc Am J.* 2016; 80: 215–226.

[13] Congreves, K.A.; Hayes, A.; Verhallen, E.A. and Van, Eerd L.L. (2015). Long-term impact of tillage and crop rotation on soil health at four temperate agroecosystems. *Soil Till Res.,* 2015; 152: 17–28.

[14] E.U. Mission Board Soil Health and Food. 2020. *Caring for Soil is Caring for Life.* 2020 Jun; European Commission; retrieved from https://op.europa.eu/en/web/eu-law-and-publications/publication-detail/-/publication/32d5d312-b689-11ea-bb7a-01aa75ed71a1.

[15] Evans, A.E.; Mateo-Sagasta, J.; Qadir, M.; Boelee, E. and Ippolito, A. (2019). Agricultural water pollution: key knowledge gaps and research needs. *Curr Opinion Environ Sustain.* 2019; 36: 20–27.

[16] Gretchen, C. Daily; Susan, Alexander; Paul, R. Ehrlich; Larry, Goulder; Jane, Lubchenco; Pamela, A. Matson; Harold, A. Mooney; Sandra, Postel; Stephen, H. Schneider; David, Tilman; George, M. Woodwell (1997). Ecosystem Services: Benefits Supplied to Human Societies by Natural Ecosystems. https://www.esa.org/esa/wp-content/uploads/2013/03/issue2.pdf

[17] Hamza, M.A. and Anderson, W.K. (2005). Soil compaction in cropping systems: A review of the nature, causes and possible solutions. *Soil Till Res.* 2005; 82: 121–145.

[18] Hanson, J.R., Macalady J.L., Harris D., Scow K.M. 1999. Linking toluene degradation with specific microbial populations in soil. *Appl Environ Microbiol.* 1999;65:5403–5408.

[19] Hussein, H.S. and Brasel, J.M. (2001). Toxicity, metabolism, and impact of mycotoxins on humans and animals. *Toxicology,* 2001; 167: 101–134.

[20] Idowu, O.J.; van Es, H.M.; Abawi, G.S.; Wolfe, D.W.; Schindelbeck, R.R.; Moebius-Clune, B.N. and Gugino, B.K. (2009). Use of an integrative soil health test for evaluation of soil management impacts. *Renew Agric Food Syst.* 2009; 24: 214–224.

[21] Jacoby, R., Peukert, M., Succurro, A., Koprivova, A. and Kopriva, S. (2017). The role of soil microorganisms in plant mineral nutrition—current knowledge and future directions. *Front Plant Sci.* 2017;8:1617.

[22] Jenkinson D.S. (1991). The Rothamsted long-term experiments: Are they still of use? *Agron J.* 1991; 83: 2–10.

[23] Jian, J., Du, X. and Stewart, R.D. (2020). A database for global soil health assessment. *Scientific Data.,* 2020; 7: 1–8.

[24] Kadam, A.M., Oza, G.H., Nemade, P.D. and Shankar, H.S. (2008). Pathogen removal from municipal wastewater in constructed soil filter. *Ecol Engin.* 2008; 33: 37–44.

[25] Karlen, D.L., Veum, K.S., Sudduth, K.A., Obrycki, J.F. and Nunes, M.R. (2019). Soil health assessment: Past accomplishments, current activities, and future opportunities. *Soil Till Res.* 2019;195:104365.

[26] Lal, R. Soil carbon sequestration impacts on global climate change and food security. *Science,* 2004; 304: 1623–1627.

[27] Lamichhane, S., Krishna, K.B. and Sarukkalige, R. (2016). Polycyclic aromatic hydrocarbons (PAHs) removal by sorption: a review. *Chemosphere,* 2016; 148: 336–353.

[28] Laurenson. G., Laurenson. S., Bolan. N. and Beecham, S. and Clark, I. (2013). The role of bioretention systems in the treatment of stormwater. *Adv Agron.,* 2013; 120: 223–274.

[29] Lehmann, J., Bossio, D.A., Kögel-Knabner, I. (2020). The concept and future prospects of soil health. *Nat Rev Earth Environ* 1, 544–553 (2020). https://doi.org/10.1038/s43017-020-0080-8

[30] Li, G., Sun, G.X., Ren, Y., Luo, X.S. and Zhu, Y.G. (2018). Urban soil and human health: a review. *Eur J Soil Sci.,* 2018; 69: 196–215.

[31] Lilburne, L.R., Hewitt, A.E., Sparling, G.P. and Selvarajah, N. (2002). Soil quality in New Zealand: policy and the science response. *J Environ Qual.,* 2002; 31: 1768–1773.

[32] Ling, L.L.; Tanja, Schneider; Aaron, J. Peoples; Amy, L. Spoering; Ina, Engels; Brian, P. Conlon; Anna, Mueller; Till, F. Schäberle; Dallas, E. Hughes; Slava, Epstein; Michael, Jones; Linos, Lazarides; Victoria, A.

Steadman; Douglas, R. Cohen; Cintia, R. Felix; K. Ashley, Fetterman; William, P. Millett; Anthony, G. Nitti; Ashley, M. Zullo; Chao, Chen and Kim, Lewis (2015). A new antibiotic kills pathogens without detectable resistance. *Nature,* 2015; 517: 455–459.

[33] Norris, C.E. and Congreves, K.A. (2018). Alternative management practices improve soil health indices in intensive vegetable cropping systems: A review. *Front Environ Sci.,* 2018; 6: 50.

[34] O' Dell, R.E. and Claassen, V.P. (2006). Vertical distribution of organic amendment influences the rooting depth of revegetation species on barren, subgrade serpentine substrate. *Plant Soil,* 2006; 285: 19–29.

[35] Ogle, S.M., Swan, A. and Paustian, K. (2012). No-till management impacts on crop productivity, carbon input and soil carbon sequestration. *Agric Ecosyst Environ.,* 2012; 149: 37–49.

[36] Oliver, M.A. and Gregory, P.J. (2015). Soil, food security and human health: a review. *Europ J. Soil Sci.,* 2015; 66: 257–276.

[37] Paustian, K.; Johannes, Lehmann; Stephen, Ogle; David, Reay; Robertson, G. Philip and Pete, Smith (2016). Climate-smart soils. *Nature,* 2016; 532: 49–57.

[38] Pittelkow, C.M.; Xinqiang, Liang; Bruce, A. Linquist; Kees, Jan van Groenigen; Juhwan, Lee; Mark, E. Lundy; Natasja, van Gestel; Johan, Six; Rodney, T. Venterea and Chris, van Kessel (2015). Productivity limits and potentials of the principles of conservation agriculture. *Nature,* 2015; 517: 365–368.

[39] Rillig, M.C., Lehmann, A., Lehmann, J., Camenzind, T. and Rauh, C. Soil biodiversity effects from field to fork. *Trends Plant Sci.,* 2018; 23: 17–24.

[40] Schatz, A., Bugle, E. and Waksman, S.A. (1944). Streptomycin, a substance exhibiting antibiotic activity against gram-positive and gram-negative bacteria. *Proc Soc Exper Biol Medicine,* 1944; 55: 66–69.

[41] Schlatter, D., Kinkel, L., Thomashow, L., Weller, D. and Paulitz, T. (2017). Disease suppressive soils: new insights from the soil microbiome. *Phytopath,* 2017; 107: 1284–1297.

[42] Sekercioglu, C.H. (2010). Ecosystem functions and services. https://doi.org/10.1093/acprof:oso/9780199554232.003.0004

[43] Sokol, N.W., Slessarev, E., Marschmann, G.L., Nicolas, A., Blazewicz, S.J. and Brodie, E.L. (2022). Life and death in the soil microbiome: How ecological processes influence biogeochemistry. *Nat Rev Microbiol* (2022) 20(7):415–30. doi: 10.1038/s41579-022-00695-z.

[44] Stott, D.E. (2019). Recommended Soil Health Indicators and Associated Laboratory Procedures. Soil Health Technical Note No. 450–03. U.S. Department of Agriculture. *Natural Resources Conservation Service.* 2019.

[45] Tournebize, J., Chaumont, C. and Mander, Ü. Implications for constructed wetlands to mitigate nitrate and pesticide pollution in agricultural drained watersheds. *Ecol Engin.,* 2017; 103: 415–425.

[46] Veresoglou, S.D., Halley, J.M. and Rillig, M.C. (2015). Extinction risk of soil biota. *Nature Commun,* 2015; 6: 8862.

[47] Wagg, C., Bender, S.F., Widmer, F., van der Heijden, M.G. (2014). Soil biodiversity and soil community composition determine ecosystem multi-functionality. *Proc Nat Acad Sci.,* 2014; 111: 5266–5270.

[48] Wall, D.H, Nielsen, U.N. and Six, J. (2015). Soil biodiversity and human health. *Nature,* 2015; 528: 69–76.

[49] Welch, R.M. and Graham, R.D. (2004). Breeding for micronutrients in staple food crops from a human nutrition perspective. *J. Exper Bot.,* 2004; 55: 353–364.

[50] Wood, S.A., Tirfessa, D. and Baudron, F. (2018). Soil organic matter underlies crop nutritional quality and productivity in smallholder agriculture. *Agric Ecosyst Environ,* 2018; 266: 100–108.

[51] Zimnicki, T.; Timothy, Boring; Grey, Evenson; Margaret, Kalcic; Douglas, L. Karlen; Robyn, S. Wilson; Yao, Zhang and Jennifer, Blesh (2020). On quantifying water quality benefits of healthy soils. *BioScience,* 2020; 70: 343–352.

Author Index